## ROUTLEDGE LIBRARY EDTIONS: GLOBAL TRANSPORT PLANNING

Volume 1

# TRANSPORT PLANNING

# TRANSPORT PLANNING
## Vision and Practice

JOHN ADAMS

Routledge
Taylor & Francis Group

LONDON AND NEW YORK

First published in 1981 by Routledge & Kegan Paul Ltd

This edition first published in 2021
by Routledge
2 Park Square, Milton Park, Abingdon, Oxon OX14 4RN

and by Routledge
605 Third Avenue, New York, NY 10017

*Routledge is an imprint of the Taylor & Francis Group, an informa business*

© 1981 John Adams

*British Library Cataloguing in Publication Data*
A catalogue record for this book is available from the British Library

ISBN: 978-0-367-69870-6 (Set)
ISBN: 978-1-00-316032-8 (Set) (ebk)
ISBN: 978-0-367-72554-9 (Volume 1) (hbk)
ISBN: 978-0-367-72598-3 (Volume 1) (pbk)
ISBN: 978-1-00-315533-1 (Volume 1) (ebk)

**Publisher's Note**
The publisher has gone to great lengths to ensure the quality of this reprint but points out that some imperfections in the original copies may be apparent.

**Disclaimer**
The publisher has made every effort to trace copyright holders and would welcome correspondence from those they have been unable to trace.

# TRANSPORT PLANNING
## Vision and practice

## John Adams

*Department of Geography*
*University College London*

Routledge & Kegan Paul
London, Boston and Henley

First published in 1981
by Routledge & Kegan Paul Ltd
39 Store Street,
London WC1E 7DD,
9 Park Street,
Boston, Mass. 02108, USA, and
Broadway House,
Newtown Road,
Henley-on-Thames,
Oxon RG9 1EN
Printed in Great Britain by
T.J. Press (Padstow) Ltd, Cornwall

Library of Congress Cataloging in Publication Data

Adams, John, 1938 Aug. 13-
Transport planning, vision and practice.

Includes index.
1. Transportation planning.   2. Transportation
planning--Great Britain.   I. Title.
HE151.A3   1981          380.5' 068          81-4941
ISBN 0-7100-0844-9                          AACR2

# CONTENTS

# ACKNOWLEDGMENTS

This book represents a drawing together of material that I have published over the past ten years. It borrows bits and pieces from contributions to the following books and journals: 'Acoustics Bulletin', 'African Urban Notes', 'Architectural Design', 'Area', 'Changing London', 'Ecologist', 'Environment and Planning', 'Geographical Journal', 'Geographical Magazine', 'Geography of Population', 'Haltwhistle Quarterly', 'Industrial Marketing Management', 'Journal of the Royal Statistical Society', 'London Journal', 'Municipal Engineering', 'New Society', 'People and their Settlements', 'Resurgence', 'Science for People', 'Systems Modelling', 'The Surveyor', 'Transport Policy Tomorrow', and 'Vole'. The most substantial of these borrowings are from 'Environment and Planning' - Chapter 14 (1977, vol. 9), Chapter 16 (1972, vol. 4), and Appendix I (1974, vol. 6); 'Vole' - Chapter 13 (1979, vol. 2, no. 7), Chapter 15 (1978, vol. 1, no. 10), Appendix III (1979, vol. 2, no. 10), and Appendix II (1977, vol. 1, no. 1); and 'Changing London' (University Tutorial Press, 1978) - Chapter 4. Chapters 9 and 10 include material originally prepared for an Open University statistics course.

Part II, which deals with transport planning practice in Britain, consists mostly of material that has not been published before. This is because the material in it, when submitted to the 'appropriate' journals, has been consistently rejected. Chapter 11 on cost-benefit analysis, for example, when submitted in essentially the form in which it is presented here to the 'Journal of Transport Economics and Policy' elicited the following reply, which I quote in full:

> I am sorry to tell you that your article is not suitable for inclusion in the Journal. Thank you for submitting it.

When submitted to the 'Ecologist', it elicited a reply with which I am not inclined to quarrel:

> I do not think that any of our readers would expect a Department of the Environment cost benefit analysis to make any sense in the first place.

When submitted to 'Regional Studies' it was rejected with the following piece of advice:

> You should appreciate by now that the entire academic community operates by a system of peer group reference. Such a system is inherently conservative. If you want to get your views accepted by your peers, then you must play the game according to their rules. Personally I am sympathetic to your arguments but I think they are best carried

forward outside of academia. As a tenured lecturer you
are free to do this and indeed should do this. You should
reserve what you may regard as more dull fare for the
learned journals.

I am indebted to Teresa Filippi and Annabel Swindells for typ-
ing not only the final copy but numerous intermediate drafts, and
to the cartographic and photographic units of the geography
departments of the University of Western Ontario and University
College London for the production of the illustrations.

I have been discussing and arguing about the subject matter of
this book with many people for many years and have incurred
more debts than I can list. But I would especially like to thank
Duke Maskell who is the most stimulating arguer I know.

# PREFACE

The book's organization reflects the development of my own views and doubts about the nature of 'progress' and the extent to which it can be planned. The view that 'development' is a diffusion process is still a deeply entrenched orthodoxy among transport planners. According to this view, the developed world has what the underdeveloped world wants, and anything that increases the contact between the two will assist the transfer of attitudes, skills and capital necessary to bring the latter up to the level of the former. This now seems to me an idea that is both naive and pernicious.

In the early 1960s, as a believer in the idea, I participated in the diffusion process. As a teacher in northern Nigeria, I was a part of an extremely selective educational system whose principal effect was the inculcation of its students with an acute sense of dissatisfaction with their village origins. It was rarely a constructive dissatisfaction but one that bred a desire to escape from these origins to the cities, which had no useful employment to offer them. After two years my contract ended and I flew out.

In retrospect this experience and the subsequent experience of writing a PhD thesis on transport and communications linkages in West Africa have forced me to recognize the arrogance inherent in the diffusionist view of progress, and persuaded me that the advice of experts who do not have to live their lives amid the consequences of their advice is likely to be untrustworthy. It also persuaded me of the importance of trying to view local problems in their global context. The level of international interdependence fostered by developments in transport and communications is now so great that it cannot be safely ignored in the planning of further developments. Part I, therefore, ventures a global perspective on transport planning.

Chapter 1 examines the belief shared by most transport planners that a high level of mobility is the legitimate aspiration of all people everywhere, and that it is the transport planner's job to help them achieve it.

Chapter 2 relates this belief to more general theories of progress and economic development that are embodied in what is termed a 'transitionalist' view of history. History, according to this view, is the story of mankind's transition from a state of poverty and subjection to the forces of nature, to a state of universal affluence and control over the forces of nature. It is a story of economic transition, demographic transition, urban transition, and *mobility transition*. All are essential aspects of the same global transition process, but it is the mobility transition that is responsible for involving distant parts of the world ever more closely in each other's affairs, and which makes it increasingly necessary to see transport problems in their global perspective.

Chapter 3 looks at the relationship between the mobility transition and the urban transition and identifies some intractable problems that these transitions jointly have created.

Chapter 4 examines transport problems in London. These problems, it is argued, are typical of those of almost all large cities in developed countries that were built before the car was a widely available form of transport.

Chapter 5 notes that a mobility transition implies an energy transition. Providing the energy necessary to achieve the levels of mobility currently being planned will, it is argued, almost certainly be impossible, and would, if it were possible, entail unacceptable costs.

Chapter 6 explores the paradox that improved methods of transport and communications are widening the gulf between the rich and the poor and are resulting in a world increasingly divided against itself.

Chapter 7 examines the way in which conventional accident statistics grossly understate the importance of road accidents as a cause of death in countries with high levels of car ownership, and discusses the curious reluctance of such countries to take effective measures to reduce the accident toll.

Societies in the grip of simplistic visions of progress can rationalize acts of incredible self-destructiveness. Cargo cults present an intriguing example. There exist in Melanesia groups of primitive people whose bizarre irrationality has captured the imagination of anthropologists from developed countries. They are Messianic religious sects whose origin is generally attributed to the psychological impact on a technologically backward people of the huge quantities of cargo disgorged from the ships and aircraft of the occupying forces during the Second World War. The millennium, according to the cosmic view of these cults, will be associated with an unlimited abundance of cargo. It is a demanding faith. In order to demonstrate their worthiness as recipients of the cargo, cult members are required to engage in the wholesale destruction of traditional forms of wealth. Such behaviour seems to me not unlike that advocated in developed countries by the proponents of economic growth for ever more. The destruction of traditional forms of wealth - treasured landscapes, established communities, wildlife, and ways of human life - required by the architects of this vision is on a vastly greater scale, but in both cases society is asked to sacrifice things of present value for the promise of highly dubious future benefits.

Part II is devoted to an examination of the rationalization rituals of pro-growth transport planners in Britain. Transport planning has become an international craft. The methods and practices of transport planners in Britain have much in common with those of

their counterparts in numerous other countries. The lessons
drawn for Britain have, therefore, a wider relevance.

Chapter 8 describes briefly the procedural steps involved in
designing and building a new road. It discusses the importance
of the policy environment within which the steps are taken, and
attributes the troubles encountered by the planners at public
inquiries to the breakdown of the consensus about the goals of
policy.

Chapters 9 and 10 look at the crucial role played by traffic
forecasts in transport planning and argue that the forecasts have
become covert policies for which neither the forecasters nor the
policy-makers will assume responsibility.

Chapter 11 examines procedures of 'assessment'. It is argued
that the economic assessment procedures employed to make transport planning decisions are incapable of assessing the desirability
of traffic growth because the desirability of growth is embodied
in the assumptions out of which the assessment models are con-
structed.

Chapter 12 explores the reasons why 'independent' public
inquiries have so conspicuously failed to convince a great many
people of the wisdom and justice of the Department of Transport's
planning decisions, and concludes that it is because the inquiries
are not independent.

Rationalization involves the invention of acceptable explanations
for behaviour that has its origin in the subconscious. An understanding of the rationalization rituals of pro-growth transport
planning requires an understanding of its subconscious moti-
vating impulses. Part III contains an exploration of this murky
metaphysical territory.

Chapter 13 depicts transport planning as a campaign to eradicate the 'disutility' of distance, and invites the reader to contemplate the consequences of the campaign succeeding.

Chapter 14 argues that the growth of mobility and the growth of
esoteric planning technology are mutually supporting trends which
are progressively diminishing the influence that an ordinary
individual can have on the institutions that shape his life.

Chapter 15 describes economics as the study of the most
efficient means of catering to insatiable appetites and concludes
that economics is unlikely to be helpful in the search for solutions
to problems that are the product of the ethos of economics.

Chapter 16 examines the inherently divisive nature of high
speed transport and communications and concludes that more of
it can only be bad for most of us.

# Part I

# PROBLEMS:
## a global perspective

# 1 THE LADDER OF PROGRESS

> Car ownership ... *should* increase, for personal mobility is
> what people want, and those who already have it should not
> try to pull the ladder up behind them ...(1)

That everyone is entitled tomorrow to what the most fortunate
enjoy today is a belief that has understandable appeal for politicians and electorates of all ideological hues. It is a belief that
dominates the planning of transport and communications, and to
challenge it is widely thought to be tantamount to committing political suicide. The following quotation is taken from 'Socialist
Commentary' but would be equally at home in the manifestos of
almost all political parties everywhere:

> no politician can ride roughshod over such a strong desire
> for personal or family transport. In the society of the future
> one must continue to expect that a wish for a car, or some
> similar means of personal transport, will rank almost as high
> among the necessities of life as a decent home, even for the
> poorest families.(2)

Sir Colin Buchanan, perhaps Britain's best-known and most
influential transport planner, explains why the desire for automotive personal mobility is so strong, and argues that it must not
be denied:

> I have never managed to make very much money, and for the
> most part, in my half century of motoring, I have made do
> with second-hand cars. But what an enrichment of life has
> resulted! Marvellous holidays - camping, caravanning, much
> of Europe at our disposal in a three week vacation. Short
> visits in infinite variety - to relatives and friends, to the
> sea, out into the country, to great houses, gardens, zoos
> and parks. Spur of the moment trips - it is a fine day so
> out we go .... Why cannot we be less hypocritical and admit
> that a motor car is just about the most convenient device that we
> ever invented, and that possession of it and usage in modera-
> tion is a perfectly legitimate ambition for all classes of people.(3)

The above three statements have been taken from discussions
of British transport planning problems arising from the growth of
car ownership, but the strength of the desire for increased personal mobility, and the legitimacy that is claimed for it are not
confined to Britain. A similar spirit is found in John Rae's 'The
Road and the Car in American Life':

> Transportation is essential to social progress; to be exact,
> transportation *is* social progress because it has been

3

throughout history the way in which not only goods and
services but ideas as well were exchanged among peoples....
The Road and the Car together have an enormous capacity
for promoting economic growth, raising standards of living,
and creating a good society. The challenge before us is to
implement this capacity.(4)

A belief in the existence of a ladder of progress that all classes
of people everywhere can and should climb also informs discussions about the prospects for travel by airplane and other more
exotic means. Sir Peter Masefield, former head of the British Airports Authority, is optimistic about the possibilities of extending
the ladder:

I have no doubt that the Ballistic Transport will appear in
the wake of the Space Shuttle and in the train of the astro-
nauts. Anywhere to anywhere in an hour - reclining comfortably, oblivious to accelerations or surroundings after a
pleasant knockout draught and before an instant reviver on
arrival. Such ballistic transport will not only be very quick
but also very cheap.... What is clear is that air transport
still has a vast contribution to make to the prosperity, the
happiness and the well being of mankind. Its disbenefits,
of noise and congestion, can be phased out - its benefits
enhanced.(5)

Such attitudes are, historically speaking, relatively recent.
Throughout history most people in most places have led pedestrian lives. Their settlement patterns and travelling have been,
as a consequence, very tightly constrained. Such vehicular transport as existed was powered by humans, animals or wind. The
rich had more mobility than the poor, but nobody had very much.
Mythologies abounding in advanced technologies - flying carpets,
winged chariots, seven-league boots, broomsticks and the like -
attest to a pervasive desire for more, but in technologically
unimaginative ages most people were resigned to this remaining
the prerogative of the gods. Indeed the legend of Icarus suggests
that the very idea of a ladder by which mere mortals might attain
such mobility was considered an impious one. Mobility, generally
speaking, was something rudimentary that people provided for
themselves rather than something planned and provided by the
state.

At a time that roughly coincides with the beginning of indus-
trialization in England there began a period of remarkable
reductions in the cost of transport and even more remarkable
increases in its speed and comfort and in the numbers who made
use of it. The achievements of the gods have been equalled and
surpassed. Concorde can fly faster than Apollo's flaming chariot
and advances in the technology of telecommunications have created
a capability for exchanging messages that far exceeds anything
ever attributed to Mercury. There have been those who have
doubted the desirability of these achievements - Thoreau writing
at the beginning of the railway age and Illich writing at the end
of it are examples - but the transport and communications history

of this period is almost invariably told as a story of economic and
social progress following in the train of technological advance. In
this story Icarus's vices of hubris and impiety have been transformed into a heroism that dares to subject the forces of nature.
It is a story of mankind becoming, if not more god-like, at least
more civilized.

Histories of transport and communications, the planning literature, and the speeches of politicians on the subject are dominated by the ladder metaphor. The historians chart the past progress of mankind's ascent, the planners and politicians project
this progress into the future. Harold Perkin, in a history of
Britain's railways, exemplifies this spirit of progress that per-
vades almost all such literature:

> All civilization depends on communication - between man and
> man, town and town, country and country, perhaps in the
> future between planet and planet.... The invention of the
> railway, next to that of the power driven factory, is Britain's
> greatest contribution to the progress of civilization, for it
> was here that the real conquest of space began. Whatever
> new frontiers of space men may conquer in the last third of
> the twentieth century, the first conquest of physical distance
> by mechanical power was *the* revolution in communications
> from which all the rest have stemmed.(6)

What has stemmed from it is an impressive amount of what advertising copy-writers for airlines call 'earth shrinking'. Figures
1.1 and 1.2 give an impression of the magnitude of the shrinkage
that has taken place as the time-distance between places has been
progressively reduced. Following the suggestion of Janelle,(7)
we can measure the rate of this shrinkage (what Janelle calls time-
space convergence) by calculating the rate at which any two
places have approached each other over time. Since the middle of
the seventeenth century London and Edinburgh have been
approaching each other at a rate of about one hour per year, and
since the early nineteenth century New York and London have
been approaching each other at about four and a half hours a year.
Janelle provides a number of other examples: since 1860 Leningrad
has been approaching Moscow at 8.35 minutes a year, Saginaw and
Detroit Michigan moved closer at a rate of 9.9 minutes a year
between 1840 and 1966, and Boston approached New York at a
speed of twenty-five minutes a year between 1800 and 1965. In all
these cases the rate of shrinkage was greatest at the beginning
of the period and had slowed to almost nothing by the end of it.
In the case of travel between London and Edinburgh, the travel
between airports and city centres and the checking-in formalities
at the airports can frequently take longer than the flight itself,
with the result that there has probably been no reduction in average travel times in the past ten years. And although Concorde has
recently reduced the flying time between New York and London by
over three hours, the addition of customs and immigration formalities and security checks to the checking-in and city centre travel
required for domestic flights can commonly make the terrestrial

parts of the journey much more time-consuming than the flight across the Atlantic. The great range in travel times shown in Figure 1.1 for the year 1760 indicates that there have also been great advances in the regularity and reliability of services since the eighteenth century.

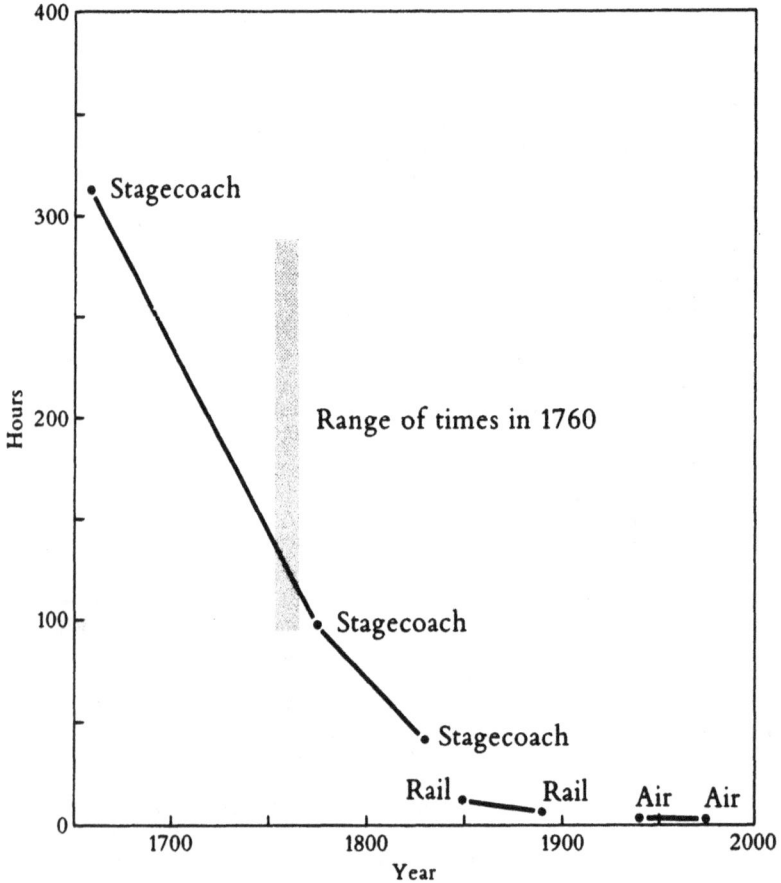

Figure 1.1   Travel times: London–Edinburgh

Sources: Perkin, H. (1970), 'The Age of the Railway', p. 26; Janelle, D. G. (1968), Central Place Development in a Time-Space Framework, 'The Professional Geographer', vol. XX, pp. 5-10; O'Dell, A. C. (1956), 'Railways and Geography', Hutchinson, pp. 179-83.

For people plugged into the global telecommunications network the process of shrinkage is now virtually complete. The time taken to exchange information between any two places in the world has been reduced to the few seconds that it takes to make the necessary electronic connections. Television and radio are now capable of informing millions of people about distant newsworthy events as they happen.

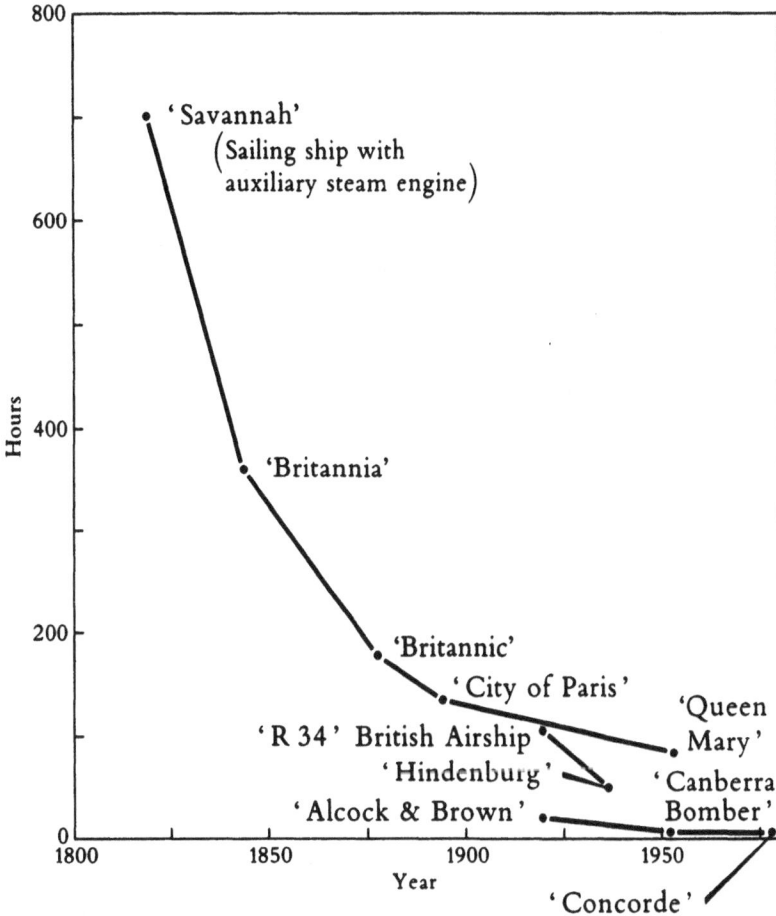

Figure 1.2   Travel times: transatlantic
Source: 'World Almanac 1978'.

There have been equally impressive reductions in the monetary cost of travel. In the middle of the eighteenth century, to travel from London to Manchester cost, depending on the style in which one travelled, the equivalent of between three months' and one year's wages for a labourer.(8) The cost today by train or car is the equivalent of less than three hours' work at the national

Table 1.1 Average distances that various types of British agricultural imports were moved, 1830-1913

| Import type | Average distance from London to regions from which each import type derived (miles) | | | | |
|---|---|---|---|---|---|
| | 1831-5 | 1856-60 | 1871-5 | 1891-5 | 1909-13 |
| Fruit and vegetables | 0 | 324 | 535 | 1150 | 1880 |
| Live animals | 0 | 630 | 870 | 3530 | 4500 |
| Butter, cheese, eggs, etc. | 262 | 530 | 1340 | 1610 | 3120 |
| Feed grains | 860 | 2030 | 2430 | 3240 | 4830 |
| Flax and seeds | 1520 | 3250 | 2770 | 4080 | 3900 |
| Meat and tallow | 2000 | 2900 | 3740 | 5050 | 6250 |
| Wheat and flour | 2430 | 2170 | 4200 | 5150 | 5950 |
| Wool and hides | 2330 | 8830 | 10 000 | 11 010 | 10 900 |
| Weighted average all above imports | 1820 | 3650 | 4300 | 5050 | 5880 |

Source: J.R. Peet (1969), The Spatial Expansion of Commercial Agriculture in the Nineteenth Century: A Von Thunen Interpretation, 'Economic Geography', vol. 45, no. 4, p. 295.

average wage. Access to electronic modes of communication has also become very much less expensive. Today, levels of telephone and television ownership suggest that in America economic barriers to these media have been almost completely overcome - although use of the telephone still varies with income.

These reductions in the time and money costs of travel and communications have been accompanied by huge increases in the movement of goods, people and information. Internationally this growth was led by the traffic in goods rather than people. Although people accompanied the goods, it was not until well into the twentieth century that long-distance mass transport facilities were provided for people. In the few exceptions to this generalization, most notably the slave trade and emigration from Europe to America, the accommodation provided was frequently of a standard suitable for livestock, or worse. Table 1.1, describing the expansion that occurred in England's agricultural hinterland, demonstrates that the consequences of the expansion of goods traffic,

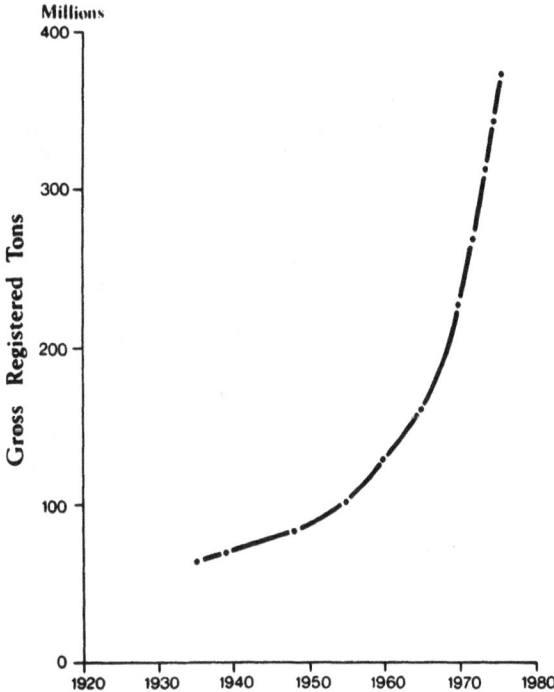

Figure 1.3   Merchant shipping: world fleet
Source: 'United Nations Statistical Yearbook'.

even in the nineteenth century, were global in their impact. In
the nineteenth century England's industrial leadership made the
country the dominant focus of an emerging world commodities
market. The simple, highly focused trading patterns of the nineteenth century have been completely overwhelmed by the increase
in the volume of trade and the complexity of trading relationships
that have taken place in the twentieth century.

Figure 1.3 plots the growth of the world's ocean freight capacity since 1935. Impressive though this growth may seem, it
appears insignificant when compared to the potential for further
growth. Indicative of the possibilities for international goods
traffic in the future if affluence is complemented by cheap transport, is Saudi Arabia's solution to the problem of providing school
meals for its rapidly expanding school system; in 1978, 200 000
school meals a day were flown in from Paris.(9)

Figures 1.4 to 1.7 illustrate, on a global scale, other salient
features of the increase in transport and communications. While
the largest absolute reductions in the cost of travel were achieved
in the nineteenth century, the largest absolute increases in the
volume of traffic have occurred since 1950. The relationship between the cost of travel and the quantity of travel appears to display the characteristics of the common demand curve of economic
text books (Figure 1.8). The explosive growth rates of global
mobility indices since the Second World War would seem to suggest
that the world as a whole is sliding towards the flat bottom part
of the curve where further small decreases in the cost of mobility
will result in extremely large increases in the amount of traffic.

The graphs on Figures 1.3 to 1.7, rising as they do almost ver-
tically off the page, provoke the obvious questions 'Where will it
all end?' The top of the mobility ladder disappears into the clouds
of science fiction. Recently, Robert Salter, a physicist with the
Rand Corporation, proposed in a paper to the American Association
for the Advancement of Science that a 'planetrain' riding a mag-
netic wave in a vacuum tube could achieve speeds of 22 500 kilometres (14 000 miles) per hour and reduce the journey time between New York and Los Angeles to twenty-one minutes.(10) Its
cost he estimated at over 250 billion dollars. Given a free scientific rein and plenty of money there appears to be no limit to the
'progress' to which some aspire. In 'The Next Ten Thousand
Years', (Coronet, 1976), which has an introductory chapter by
astronomer Patrick Moore proclaiming it a work of serious scientific speculation, Adrian Berry asserts 'there are no limits to
growth'. He extrapolates twentieth-century exponential trends in
economic growth, energy consumption, and mobility into an indefinite future in which the energy and materials required to sustain this growth will be obtained by strip-mining, and even
rearranging, distant galaxies. He refuses to accept even the
speed of light as an ultimate constraint on human mobility. He
speculates that journeys of millions of light years might be accomplished instantaneously by means of short-cuts through 'super-
space'.

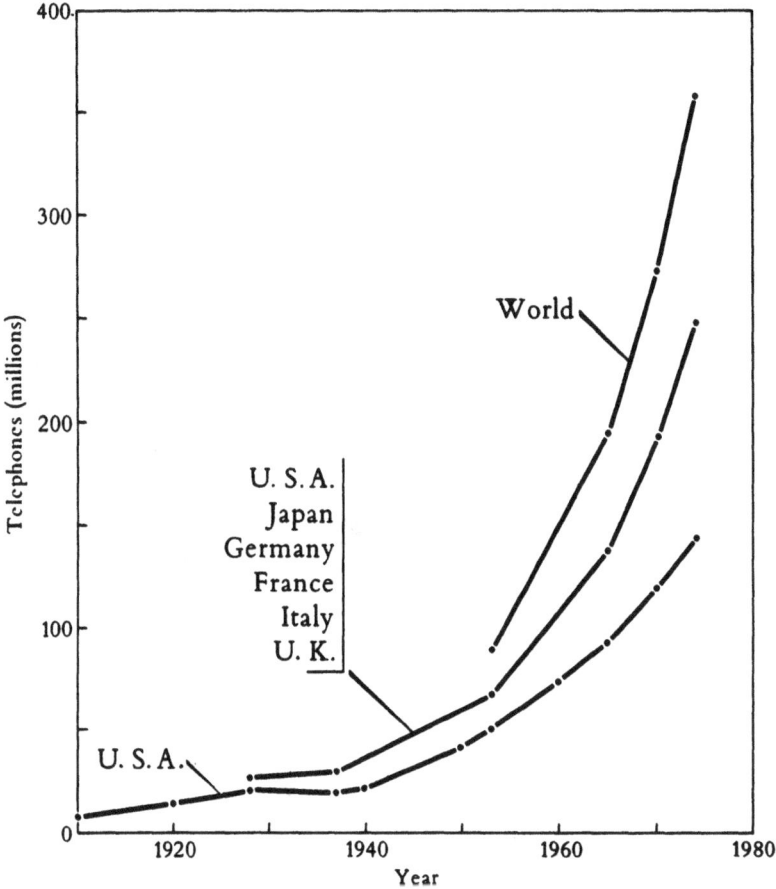

Figure 1.4   Telephones
Sources: 'United Nations Statistical Yearbooks'; 'Statistical History of the United States from Colonial Times to 1970', Rl.

It is possible, however, to be less speculative about what would be involved on terra firma if the whole world were to catch up with the current mobility leaders. For purposes of describing trends in mobility over time and differences between various parts of the world at any moment in time it would be extremely useful to have a single summary index of mobility. Because of the variety of modes by which and purposes for which people travel, and also because of the great variability in the quantity and quality of transport and communications statistics throughout the world,

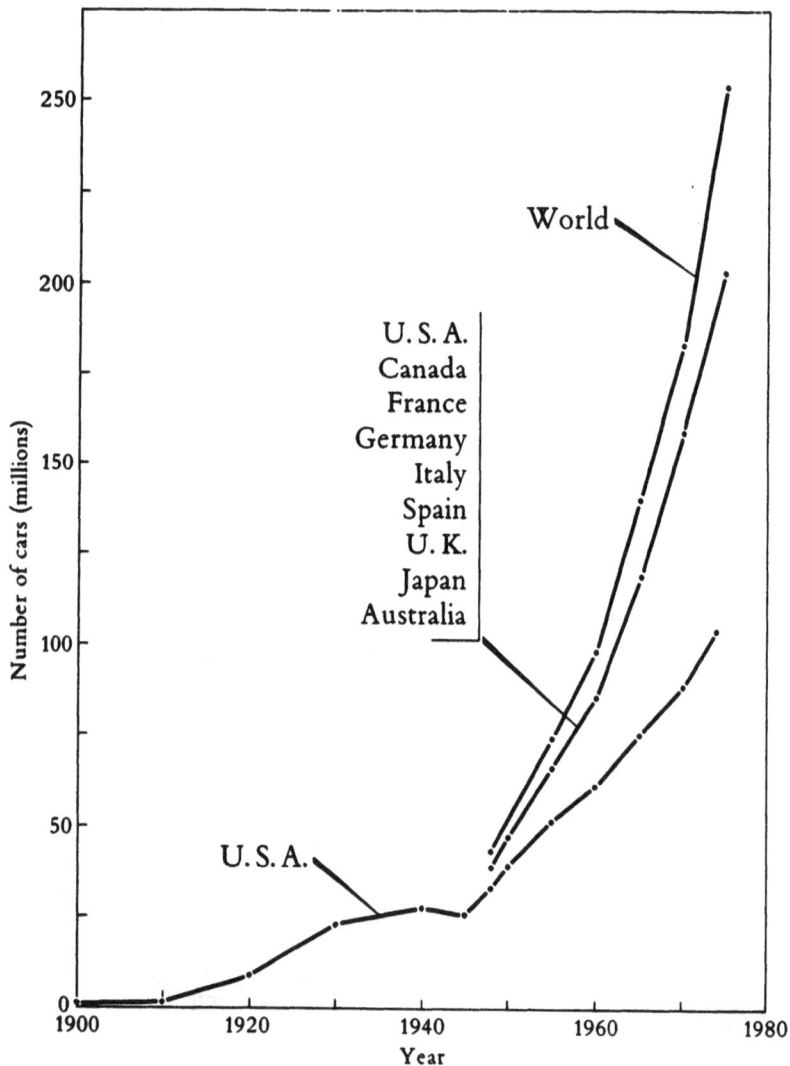

**Figure 1.5    Cars**

Sources: 'United Nations Statistical Yearbooks'; 'Statistical
    History of the United States from Colonial Times to
    1970', Q153.

such an ideal index is unlikely ever to be available. But a useful
indication of a given society's level of mobility that can be roughly
calculated for most societies is the number of kilometres the average member of it travels in a day. Estimates of the amount of
travel that took place in pre-industrial societies without mechanical means of transport are extremely rough. But in non-nomadic
societies travel outside one's village was, for most people, very
infrequent, and the dimensions of the settlements in which they
lived suggest that average daily travel per person, including
women and children and old people, was probably a small fraction
of a mile a day. In 'Central Places in Southern Germany', (11)
for example, Christaller found that the hinterlands of agricultural
villages, the lowest order central places, tended with a remarkable consistency to have radii between 4-4.8 kilometres (2½-3 miles
and this for most people most of the time would have been an outer
limit to their travels. In a recent article in the 'Journal of Transport History', G. H. Martin, in support of his argument that
there was much more travel in the middle ages than is commonly
supposed, adduced as evidence the fact that in the whole of the
year 1381 there were more than 3000 carts recorded as having
been charged tolls for using London Bridge, then London's only

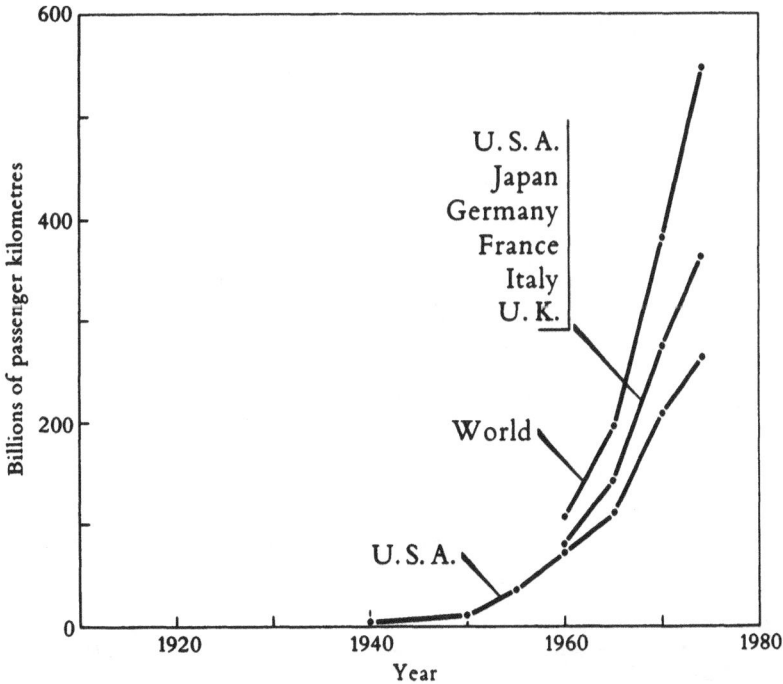

Figure 1.6   Scheduled air services

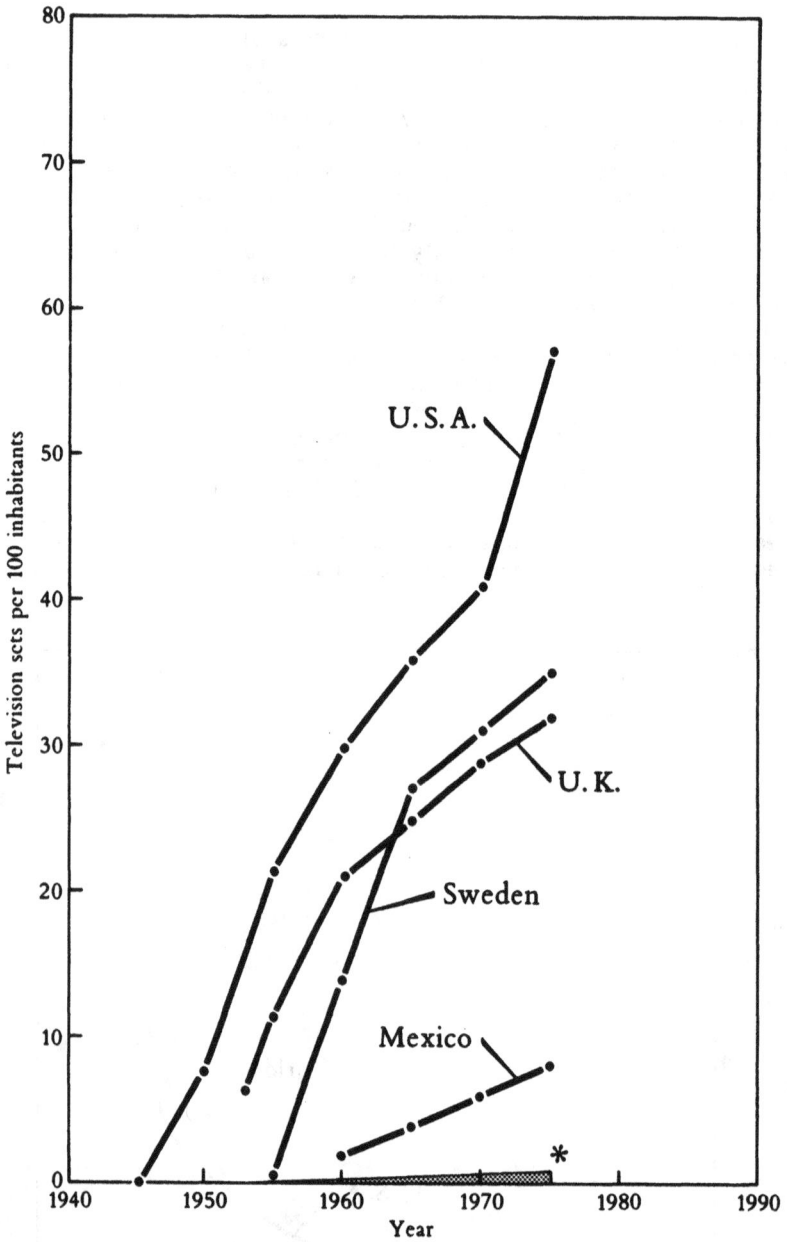

Figure 1.7   Television
*Represents 69 countries with less than 1 television for every 100 inhabitants

Sources: 'World Almanac 1978'; 'Statistical Abstract of the US'.

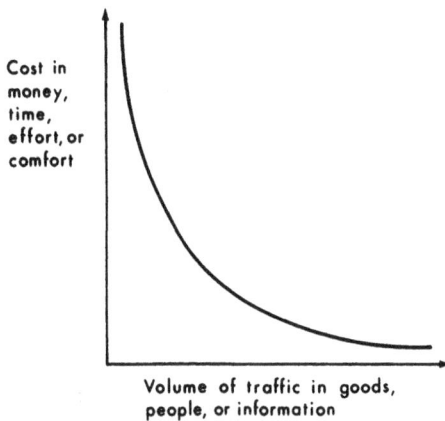

Figure 1.8   Demand for mobility

bridge.(12) By 1962, despite the restraining influence of congestion, more than ten times this number of vehicles crossed
London Bridge every day, and the total number of vehicles cros-
sing the Thames daily by London's twenty-one bridges and two
tunnels was over 600 000.(13)

On a typical day in Britain in 1973 the average citizen travelled
by bus, train, car, van, taxi, motor-cycle and airplane approximately 24 kilometres (15 miles). In the same year the average
American travelled 45 kilometres (28 miles). They also covered a
lot of ground electronically. Britain averages almost one phone
call per person per day and the United States more than two. The
average time spent watching television is about three hours per
person a day in Britain and over four in the United States. Of the
kilometres travelled physically in Britain 75 per cent were travelled in cars and in the United States 83 per cent (see Table 6.1,
p. 83). In most countries the average number of kilometres
travelled by each car in a year remains fairly constant; in Britain
and the United States it is between 9000 and 10 000. Thus for the
purpose of making rough and ready international comparisons involving countries with less comprehensive transport statistics
than Britain and the United States the number of cars in a country is probably the single best guide to its position on the international mobility ladder (see Table 1.2, pp. 18-21).

In undisputed occupation of the highest rung of the international
ladder is the United States with about 110 000 000 cars; and in
undisputed possession of the highest rung of the American ladder
is Los Angeles with an estimated car population in 1970 of more
than 3 900 000 and now well over 4 million. The following is a list
of countries which, according to UN statistics, have fewer cars
than Los Angeles: Afghanistan, Bangladesh, Burma, China,
Kampuchea, India, Indonesia, Laos, Nepal, Pakistan, Philippines,
Vietnam, Sri Lanka, Thailand, Yemen, Angola, Benin, Botswana,

Burundi, Cape Verde, Central African Republic, Chad, Egypt,
Ethiopia, Gambia, Guinea, Kenya, Lesotho, Liberia, Madagascar,
Malawi, Mali, Mauritania, Mozambique, Niger, Nigeria, Rwanda,
Sao Tome, Senegal, Seychelles, Sierra Leone, Somalia, Sudan,
Swaziland, Togo, Uganda, Cameroon, Tanzania, Upper Volta,
Zaire, Zambia, Haiti, Honduras, Grenada, El Salvador, Bolivia,
Papua New Guinea, Tonga and Samoa. This is by no means a
complete list. As reference to Table 1.2 will confirm there are
only eight countries in the world, in addition to the United States,
that have more cars than Los Angeles. The above list contains the
fifty-nine poorest countries in the world all with per capita incomes in 1976 of less than $500 a year. In population they range
from the biggest to the smallest and collectively they contain
about 58 per cent of the world's population. *Together* they own
fewer cars than Los Angeles.

Despite the rapid rate at which a few countries are climbing
the ladder of car ownership, the number of people in the world
who have yet to get a foot on the bottom rung is larger than ever.
At the beginning of this century there were approximately 1.6
billion people who did not own cars. Today there are more than
twice that many. Despite this emphatic lack of progress in catching up with the leaders, everywhere in the world transport planners and politicians hold up the United States as the model to
emulate. At inquiries into road schemes in Britain the Department
of Transport's road planners offer the American experience, quite
explicitly, as justification for their predictions that car ownership will not cease to grow until it reaches levels already realized
in the United States. Having surveyed transport planning practices in thirty of the world's largest cities in twenty different
countries at all stages of economic development, Michael Thomson
concluded that 'most governments regard the general desire for
car ownership as irresistible'.(14)

Is it conceivable that the whole world might one day enjoy the
level of car ownership and car use that Los Angeles enjoys today?
The most common answer to this question is that it is politically
inconceivable that it might not. Michael Thomson writing about
Britain in 1971 in an article entitled Halfway to a Motorized Society
asserted 'We can confidently predict that by the end of the century almost every household that wants a car will have one. No
government is likely to do anything that substantially alters this
fact.... We are no longer just speculating about a motorized
society we are actually designing it.'(15)

Although the future growth in car traffic implied in leaving the
ladder in place and encouraging the whole world to join the present occupants of the uppermost rungs is impressive, the growth
in air traffic implicit in this levelling-up philosophy would be very
many times greater. The vast majority of the world's population
have never flown, let alone flown regularly. But it is air travel
above all other modes of travel that brings into physical proximity
those who fly a great deal and those who fly not at all. It is
international travel that demands the extension of the ladder

metaphor to the whole world. Sir Colin Buchanan now has most of
Europe 'at his disposal' for his holidays and thinks that it ought
to be at the disposal of all classes. With modern jet aircraft, and
the promise of ballistic transport to come, the whole world is at
the disposal of all those who can afford the price of a ticket. Few
politicians can be found who will deny that a ticket is a 'legitimate
ambition for all classes of people'.

The growth trends illustrated by Figures 1.3 to 1.7 have enormous economic and political momentum. They provoke questions
with no agreed answers. Can they, will they, should they continue? The answer of this book is no, no, no!

## REFERENCES AND NOTES

1 Department of the Environment (1976), 'Transport Policy: A
  Consultation Document', HMSO.
2 'Socialist Commentary', April 1975, a special issue on transport
  policy.
3 Buchanan, C. (1973), Some Thoughts about the Motor Car,
  'Traffic Engineering and Control', July.
4 Rae, J. (1971), 'The Road and the Car in American Life', MIT
  Press.
5 Masefield, P. (1973), The Air Transport Scene: Present Problems - Future Prospects, 'The Three Banks Review', June,
  no. 98.
6 Perkin, H. (1970), 'The Age of the Railway', Routledge &
  Kegan Paul, p. 11.
7 Janelle, D. G. (1968), Central Place Development in a Time-
  Space Framework, 'The Professional Geographer', vol. XX,
  pp. 5-10.
8 Perkin (1970), op. cit., p. 24.
9 'Time Magazine', 22 May 1978.
10 'Toronto Globe and Mail', 14 February 1978.
11 Christaller, W. (1933), trans. C. W. Baskin, Prentice-Hall,
  1966.
12 Martin, G. H. (1976), Some Journeys by the Warden and Fel-
  lows of Merton College, Oxford, 1315-1470, 'The Journal of
  Transport History'.
13 'London Traffic Survey', London County Council, 1962, p. 17.
14 Thomson, M. (1978), 'Great Cities and Their Traffic', Penguin.
15 Thomson, M. (1971), Halfway to a Motorized Society, 'Lloyds
  Bank Review', October, no. 102.

*Table 1.2 Positions on the mobility ladder*

| Country | Population (millions) | Cars (thousands) | Cars per 100 inhabitants | Telephones per 100 inhabitants |
|---|---|---|---|---|
| WORLD | 4 083.0 | 258 080.0 | 6.3 | 9.6 |
| AFRICA | | | | |
| Algeria | 17.3 | 180.0 | 1.0 | 1.4 |
| Angola | 5.8 | 127.3 | 2.2 | 0.6 |
| Benin | 3.2 | 14.0 | .4 | 0.3 |
| Botswana | .7 | 3.4 | .5 | 1.2 |
| Burundi | 3.9 | 4.2 | .1 | 0.1 |
| Cape Verde | .3 | 2.7 | .9 | 0.5 |
| Central African Rep. | 2.6 | 9.1 | .4 | 0.2 |
| Chad | 4.1 | 5.8 | .1 | 0.2 |
| Congo | 1.4 | 19.0 | 1.3 | 0.8 |
| Egypt | 38.0 | 215.5 | .6 | 1.4 |
| Ethiopia | 28.7 | 41.0 | .1 | 0.3 |
| Gabon | .5 | 10.1 | 2.0 | 1.2 |
| Gambia | .5 | 3.0 | .6 | 0.5 |
| Ghana | 10.3 | 55.5 | .5 | 0.6 |
| Guinea | 4.5 | 10.2 | .2 | 0.2 |
| Ivory Coast | 5.0 | 90.5 | 1.8 | 0.9 |
| Kenya | 13.9 | 130.9 | 0.9 | 0.9 |
| Lesotho | 1.0 | 4.6 | 0.5 | 0.3 |
| Liberia | 1.8 | 12.1 | 0.7 | 0.3 |
| Libya | 2.4 | 263.1 | 11.0 | 2.1 |
| Madagascar | 8.3 | 55.0 | 0.7 | 0.4 |
| Malawi | 5.2 | 11.2 | 0.2 | 0.4 |
| Mali | 5.8 | 15.0 | 0.3 | 0.1 |
| Mauritania | 1.3 | 4.4 | 0.3 | n.a. |
| Mauritius | .9 | 17.8 | 2.0 | 2.9 |
| Morocco | 17.8 | 320.1 | 1.8 | 1.0 |
| Mozambique | 9.4 | 89.3 | 1.0 | 0.6 |
| Niger | 4.7 | 8.6 | 0.2 | 0.1 |
| Nigeria | 64.8 | 150.0 | 0.2 | 0.2 |
| Rwanda | 4.3 | 6.5 | 0.2 | 0.1 |
| Sao Tome | .08 | 1.6 | 2.0 | n.a. |
| Senegal | 5.1 | 44.8 | 0.9 | 0.9 |
| Seychelles | 0.6 | 2.5 | 4.0 | 5.6 |
| Sierra Leone | 3.1 | 14.8 | 0.5 | 0.4 |
| Somalia | 3.3 | 8.0 | 0.2 | 0.2 |
| South Africa | 26.1 | 2 117.0 | 8.1 | 7.8 |
| Southern Rhodesia | 6.5 | 180.0 | 2.8 | 2.8 |
| Sudan | 16.1 | 29.2 | 0.2 | 0.3 |
| Swaziland | 0.5 | 7.1 | 1.4 | 1.5 |
| Togo | 2.3 | 13.0 | 0.6 | 0.3 |
| Tunisia | 5.7 | 102.6 | 1.8 | 2.3 |
| Uganda | 11.9 | 27.0 | 0.2 | 0.4 |
| Cameroon | 6.5 | 39.1 | 0.6 | 0.4 |
| Tanzania | 15.6 | 39.1 | 0.3 | 0.4 |
| Upper Volta | 6.2 | 9.5 | 0.2 | 0.1 |
| Zaire | 25.6 | 84.8 | 0.3 | 0.2 |
| Zambia | 5.1 | 85.8 | 1.7 | 1.7 |

*Table 1.2 Positions on the mobility ladder*

| Country | Population (millions) | Cars (thousands) | Cars per 100 inhabitants | Telephones per 100 inhabitants |
|---|---|---|---|---|
| **NORTH AMERICA** | | | | |
| Bahamas | 0.2 | 40.1 | 20.0 | 28.0 |
| Barbados | 0.3 | 20.5 | 6.8 | 17.2 |
| Canada | 23.0 | 8 472.2 | 36.8 | 57.2 |
| Costa Rica | 2.0 | 55.1 | 2.8 | 5.6 |
| Cuba | 9.3 | 70.0 | 0.8 | 3.2 |
| Dominican Rep. | 4.8 | 71.5 | 1.5 | 2.4 |
| El Salvador | 4.0 | 41.0 | 1.0 | 1.4 |
| Grenada | 0.1 | 3.8 | 3.8 | 4.5 |
| Guatemala | 6.3 | 76.1 | 1.2 | 1.0 |
| Haiti | 4.7 | 11.7 | 0.2 | 0.2 |
| Honduras | 3.1 | 14.7 | 0.5 | 0.7 |
| Jamaica | 2.1 | 86.4 | 4.1 | 5.0 |
| Mexico | 62.3 | 2 400.9 | 3.9 | 2.8 |
| Nicaragua | 2.2 | 32.0 | 1.5 | 1.0 |
| Panama | 1.7 | 62.6 | 3.7 | 8.5 |
| Trinidad | 1.1 | 101.3 | 9.2 | 6.0 |
| United States | 216.0 | 106 712.5 | 49.4 | 69.5 |
| **SOUTH AMERICA** | | | | |
| Argentina | 25.8 | 2 027.5 | 7.9 | 7.8 |
| Bolivia | 5.8 | 9.1 | 0.2 | 0.9 |
| Brazil | 109.2 | 3 679.3 | 3.4 | 3.1 |
| Chile | 10.5 | 236.8 | 2.3 | 4.5 |
| Colombia | 24.4 | 326.9 | 1.3 | 5.5 |
| Ecuador | 7.3 | 43.6 | 0.6 | 2.7 |
| Guyana | 0.8 | 25.5 | 3.2 | 2.6 |
| Paraguay | 2.7 | 16.0 | 0.6 | 1.4 |
| Peru | 16.1 | 266.9 | 1.7 | 2.1 |
| Surinam | 0.4 | 21.5 | 5.4 | 4.2 |
| Uruguay | 3.1 | 151.6 | 4.9 | 9.0 |
| Venezuela | 12.4 | 601.1 | 4.8 | 5.3 |
| **ASIA** | | | | |
| Afghanistan | 19.8 | 38.4 | 0.2 | 0.2 |
| Bahrain | .26 | 23.3 | 9.0 | 10.0 |
| Bangladesh | 76.8 | 31.7 | .04 | 0.1 |
| Burma | 30.8 | 36.3 | 0.1 | 0.1 |
| China[a] | 852.0 | 50.0 | .01 | n.a. |
| Kampuchea | 8.4 | 27.2 | 0.3 | 11.2 |
| India | 610.0 | 756.5 | 0.1 | 0.3 |
| Indonesia | 139.6 | 383.1 | 0.3 | 0.2 |
| Iran | 33.9 | 589.2 | 1.7 | 2.0 |
| Iraq | 11.5 | 80.1 | 0.7 | 1.7 |
| Israel | 3.5 | 284.0 | 8.1 | 23.1 |
| Japan | 112.7 | 17 236.0 | 15.3 | 40.5 |
| Jordan | 2.8 | 33.1 | 1.2 | 1.6 |
| South Korea | 35.9 | 84.2 | 0.2 | 4.0 |

*Table 1.2  Positions on the mobility ladder*

| Country | Population (millions) | Cars (thousands) | Cars per 100 inhabitants | Telephones per 100 inhabitants |
|---|---|---|---|---|
| Kuwait | 1.0 | 203.7 | 20.4 | 12.3 |
| Laos | 3.4 | 14.1 | 0.4 | 0.2 |
| Lebanon | 3.0 | 220.2 | 7.3 | 7.7 |
| Malaysia | 12.3 | 37.4 | 3.6 | 2.5 |
| Nepal | 12.9 | 4.0 | .03 | 0.1 |
| Pakistan | 72.4 | 177.3 | .2 | 0.3 |
| Philippines | 43.8 | 362.5 | .8 | 1.2 |
| Saudi Arabia | 9.2 | 59.4 | .6 | 1.0 |
| Singapore | 2.3 | 149.0 | 6.5 | 12.9 |
| Vietnam(b) | 46.5 | 70.0 | .2 | 0.3 |
| Sri Lanka | 14.3 | 91.7 | .6 | 0.5 |
| Syria | 7.6 | 50.2 | .6 | 2.1 |
| Thailand | 43.0 | 286.2 | .7 | 0.7 |
| Turkey | 40.2 | 303.8 | .8 | 2.5 |
| Yemen, Democratic | 1.8 | 10.6 | .6 | 0.6 |
| **EUROPE** | | | | |
| Austria | 7.5 | 1 720.7 | 22.9 | 28.1 |
| Belgium | 9.9 | 2 613.9 | 26.4 | 28.5 |
| Czechoslovakia | 14.9 | 1 505.1 | 10.1 | 17.6 |
| Denmark | 5.1 | 1 300.0 | 25.5 | 45.4 |
| Finland | 4.7 | 996.3 | 21.2 | 38.9 |
| France | 52.9 | 15 300.0 | 28.9 | 26.2 |
| East Germany | 16.8 | 1 880.5 | 11.2 | 15.2 |
| West Germany | 61.5 | 17 898.3 | 29.1 | 31.7 |
| Greece | 9.2 | 439.1 | 4.8 | 22.1 |
| Hungary | 10.6 | 579.9 | 5.5 | 9.9 |
| Iceland | .2 | 63.9 | 32.0 | 41.7 |
| Ireland | 3.2 | 515.6 | 16.1 | 14.1 |
| Italy | 56.2 | 14 295.0 | 25.4 | 25.9 |
| Luxembourg | .4 | 140.1 | 35.0 | 41.1 |
| Malta | .3 | 54.0 | 18.0 | 16.3 |
| Netherlands | 13.7 | 3 399.0 | 24.8 | 36.8 |
| Norway | 4.0 | 953.7 | 23.8 | 35.0 |
| Poland | 34.4 | 1 077.7 | 3.1 | 7.5 |
| Portugal | 9.5 | 937.0 | 9.9 | 11.3 |
| Romania | 21.5 | 45.1 | .2 | 5.1 |
| Spain | 36.0 | 4 806.8 | 13.3 | 22.0 |
| Sweden | 8.2 | 2 760.0 | 33.7 | 66.1 |
| Switzerland | 6.4 | 1 723.0 | 26.9 | 61.1 |
| United Kingdom | 55.9 | 14 263.0 | 25.5 | 37.9 |
| Yugoslavia | 21.6 | 1 536.7 | 7.1 | 6.1 |
| **OCEANIA** | | | | |
| Australia | 13.6 | 5 012.3 | 36.9 | 39.0 |
| Fiji | .6 | 21.5 | 3.6 | 5.0 |
| New Zealand | 3.1 | 1 167.6 | 37.6 | 50.2 |
| Papua New Guinea | 2.8 | 17.3 | .6 | 1.3 |

*Table 1.2  Positions on the mobility ladder*

| Country | Population (millions) | Cars (thousands) | Cars per 100 inhabitants | Telephones per 100 inhabitants |
|---|---|---|---|---|
| Tonga | .09 | 1.0 | 1.1 | n.a. |
| Samoa | .16 | 2.0 | 1.25 | 2.1 |
| USSR | 257.9 | 3 000.0 | 1.2 | 6.6 |
| Los Angeles urbanized area[c] | 8.35 | 3 977.6 | 48.0 | n.a. |

Source: The principal source for the above table is the 'United Nations Statistical Yearbook 1976'. The figures are for the most recent year given, in most cases 1975

(a) The number of cars in China is a guess. The UN does not give a figure. 'Time Magazine', 13 March 1978, puts it at one car for every 10 000 people, or about 85 000. 'The International Petroleum Encyclopedia 1974' guesses 28 900 and 'The World Almanac 1978', 30 000.

(b) South Vietnam only.

(c) Source: 'Transportation Planning Data for Urbanized Areas; 1970', US Department of Transport, 1973.

# 2 THEORIES OF TRANSITION

The view that the trends described in Chapter 1 can, will and should continue is firmly embedded in all Marxist and capitalist theories of economic development. Whether or not these trends represent 'progress' is not a question of mere historical interest. The purpose of theorizing about development is to identify its causes and inhibitors in order to encourage the former and discourage the latter.

Mobility has increased not only within countries but among countries and has transformed the question of whether or not its causes should be further encouraged into one that the whole world must answer. Not only has there been a dramatic increase in the numbers of people visiting, or intruding upon, the territory of nations not their own, there has also been a great increase in the dependence of most nations on the import of raw materials, especially energy, to sustain the levels of mobility that they currently enjoy, or aspire to. This growing global interdependence has led to an increasingly close scrutiny by the less well-off nations of the way in which the material standard of living of the better-off impinges on their own aspirations for development. It has played an important part in the creation of a universal sense of entitlement to affluence, and in the standardization of the vision of affluence. And it has made it increasingly difficult to dismiss awkward questions such as 'What would the world be like if all countries everywhere were to achieve the standards of consumption and mobility currently enjoyed by Californians?'

STAGES OF GROWTH

In 'The Stages of Economic Growth; A Non-Communist Manifesto' (1) by W. W. Rostow, arguably the most influential economic historian since Marx, the object of this universal sense of entitlement achieves the status of mankind's manifest destiny. Rostow's primary concern appears to be not with the difficulties in the path of economic progress, but with convincing the sceptic that the world is progressing inexorably towards its desired end, 'the stage of high mass consumption'. Although he hesitates to predict the future course of development beyond that stage, he argues that the course followed by developed countries that have already reached this stage serves as a model that other countries will follow. This course he summarizes as follows:

It is possible to identify all societies, in their economic

dimensions, as lying within one of five categories: the
traditional society, the preconditions for take-off, the take-
off, the drive to maturity, and the age of high mass consumption.(2)
And while he objects to being labelled a determinist he offers his
interpretation of history as evidence for almost inevitable progress:

It is as sure as anything can be that, barring a global catastrophe, the societies of the underdeveloped areas will
move through the transitional processes and establish the
preconditions for take-off into economic growth and modernization. And they will then continue the process of
sustained growth and move on to maturity; that is to the
stage when their societies are so structured that they can
bring to bear on their resources the full capabilities of
modern technology.(3)

It was in Western Europe in the eighteenth century during the
establishment of the preconditions for take-off that a part of mankind experienced for the first time what Rostow calls 'the blessings
and choices opened up by the march of compound interest'. And it
was first along improved tracks, then roads, canals, railways, sea
lanes and finally airways that the marching took place. Every step
up towards the stage of high mass consumption coincides in his
model with advances in transport that permitted and encouraged
more trade and travel over greater distances. Whatever a country's political ideology, such increases are necessary for increases
in specialization in production which, in turn, are necessary for
increasing per capita productivity. The arrival of a country at the
final stage is signalled, according to Rostow, by the rapid spread
of car ownership: 'Historically ... the decisive element has been
the cheap mass automobile with its quite revolutionary effects -
social as well as economic - on the life and expectations of
society.'

## GEOGRAPHICAL MODELS

The process described by Rostow is a diffusion process. According to most geographical accounts, the modern process of economic development outside Europe began with the arrival of
Europeans. Although 'traditional' economic activities existed prior
to this, the original patterns were either obliterated or transformed beyond recognition by the European influence. During the
establishment of the preconditions for take-off, and subsequently,
the modern exchange sectors of the economies of most of the non-
European world were dominated by Europeans. Although many
indigenous people were employed in this sector, the pace and
direction of change were controlled by entrepreneurs from, or
governments of, the European powers.

That this change was almost wholly beneficial is not doubted by
development economists such as P. T. Bauer:

In many areas this progress has meant the suppression of

slavery and tribal warfare and the disappearance of famine and of the worst epidemic and endemic diseases. It has meant the development of communications, the replacement of local self-sufficiency by the possibilities of exchange and the emergence of cities.

The fact that this progress was accompanied by the military control of many of the areas concerned does not alter his judgement:

Colonial status has not precluded the material advance of African or Asian territories which became colonies in the nineteenth century. Many of these territories made rapid economic progress between the second half of the nineteenth century when they became colonies, and the middle of the twentieth century, when most of them became independent.(4)

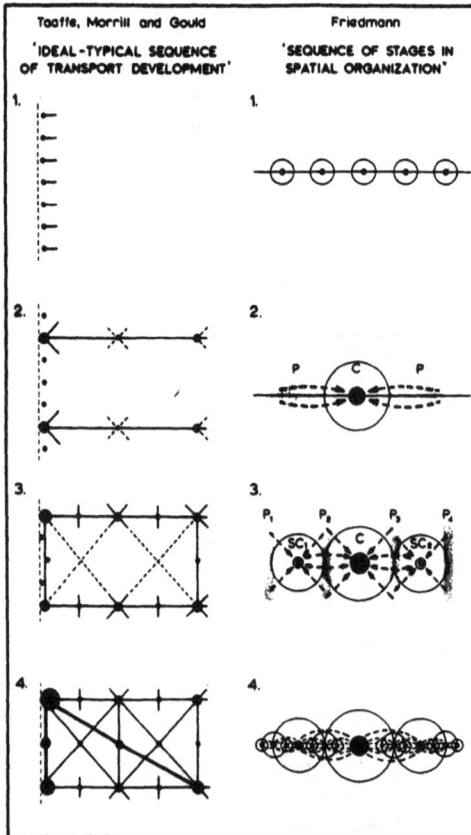

Figure 2.1 Development sequences

And in areas where progress has not been made, it is in spite of, not because of, the European influence: 'What holds back the underdeveloped countries is the people living there'.(5)

But he concedes that progress has sometimes required that some people swallow a bit of pride: 'Colonial status may well irritate or even humiliate certain sections of the population. But it does not follow that this status obstructs material advance'.(6)

Attempts have been made to represent the geographical dimension of Rostow's stages of growth sequence by Taaffe, Morrill and Gould in 1963 (7) and Friedmann in 1966.(8) The models, illustrated by Figure 2.1, are basically very similar in that both depict sequences of geographical spread accompanied by increasing complexity of linkage. Friedmann was primarily concerned with the sequence of development in Latin America, while Taaffe, Morrill and Gould were concerned primarily with West Africa. Hence both models describe development sequences in which the original stimulus to development is attributed to the arrival of Europeans rather than to endogenous forces. They both adhere faithfully in their various stages to the Rostow model.

| The Taaffe-Morrill-Gould model | The Friedmann model |
|---|---|
| **Stage 1** | |
| There is a small scattering of ports along the seacoast and little lateral interconnectedness; each port has a limited hinterland. | Independent local centres and the absence of a hierarchy are typical of a pre-industrial structure; each city lies at the centre of a small regional enclave; growth possibilities are quickly exhausted; the economy tends to stagnate. |

In the first stage of the Taaffe model the ports perform a trading function. This activity is centred on European forts or trading stations that organize the collection of produce from a limited hinterland and the distribution of European-manufactured goods to this hinterland. In the Friedmann model no external linkages are mentioned; the centres are described as isolated and independent. However, in neither model is there any interaction between the centres within the geographical area encompassed by the model. Both Taaffe and Friedmann compare this stage of their models to Rostow's 'traditional society' stage.

| | |
|---|---|
| **Stage 2** | |
| Figure 2.1 combines the second and third stages of the Taaffe model. One or two ports begin to dominate while the rest decline; the dominant ports enlarge their hinterlands and develop links with the interior. | A core-periphery pattern begins to emerge; linkages develop but are mostly one way; there is a migration of entrepreneurs, intellectuals, labour and capital to the centre; the economy is virtually reduced to a single metropolitan region. |

This stage is compared by both authors to Rostow's 'take-off' stage. If the Taaffe model is divided horizontally across the middle, the changes described in the coastal regions of either one of its two parts are similar to the changes described in the Friedmann model. Since there is little interaction between the two halves of the Taaffe model the comparison is reasonable; in both models one centre dominates at the expense of its neighbours. The appearance of inland centres for the first time in the Taaffe model apparently creates a problem because it suggests an expansion of activity in the major city's peripheral region, whereas the Friedmann model suggests a decline. In the West African context this is somewhat misleading, because although a number of inland centres become parts of the hinterlands of coastal cities for the first time, most of them suffered a great decline in power relative to the coastal cities. The primary orientation of commerce, which had been previously towards the Sahara, was now towards the coast and the roles of core and periphery were reversed.

### Stage 3

Stage 3 in Figure 2.1 combines the fourth and fifth stages of the original model. The two dominant ports have been linked; many more internal connections have been established, and a few major centres in the interior are depicted as growing vigorously.

A single national centre remains; strong regional centres are emerging; the flow of resources to the centre has been reversed; there is greater inter-regional connectivity; problems of poverty persist in the intermetropolitan peripheries.

This stage belongs to the 'drive to maturity stage' of Rostow's model. The models depict a reversal of the earlier process of concentration at the centre. Development at this stage is viewed as a diffusion process by which stimuli to growth spread from the coast, or the core, to the interior, or the periphery. There is a marked spatial integration of the economy at this stage.

### Stage 4

Lateral interconnection continues until all ports, interior centres and main nodes are linked to the national economy; national trunk routes or 'main streets' develop between major nodes which grow at the expense of lesser nodes.

The final stage is represented by a functionally interdependent system of cities; organized complexity increases during the period of industrial maturation.

This stage corresponds in both models to Rostow's 'age of high mass consumption'. In both models one centre, the primate city, dominates, and beneath this there is a hierarchy of settlements. The integration of the space economy is complete and the centrifugal and centripetal forces are in balance.

## THE MOBILITY TRANSITION

Another prominent multi-stage model that treats development, or 'modernization', as a diffusion process is Zelinsky's 'Hypothesis of the Mobility Transition'. He states his hypothesis as follows: 'There are definite, patterned regularities in the growth of personal mobility through space-time during recent history, and these regularities comprise an essential component of the modernization process'.(9) The Zelinsky model places even more emphasis than the models of Rostow, Friedmann and Taaffe et al. on the role of transport in the development process. Oddly, in the presentation of his model he makes no mention of the earlier models. This omission probably stems from his preoccupation with the correspondence between the stages in his mobility model and those of the 'demographic transition'. All the models correspond closely.

Zelinsky distinguishes three types of mobility: migration, or residential mobility; circulation, journeys that begin and end at a home base; and communication, or electronic mobility. The stages of the mobility transition are defined in terms of changes in these three types of mobility.

*1 The Pre-modern Traditional Society*
This stage is identical to the starting point of the other models. It is characterized by very low levels of migration and circulation. It corresponds to the first, low-growth stage of the demographic transition in which high fertility levels are offset by high mortality levels.

*2 The Early Transitional Society*
This stage embraces the precondition and take-off stages of the earlier models. There is massive movement from rural areas into towns and cities and a significant increase in circulation. Substantial emigration from Europe, and frontierward movement in the areas of immigration also occur in this stage. This is the stage in the demographic transition of extremely rapid growth; fertility is very high and mortality rates decline sharply.

*3 The Late Transitional Society*
This corresponds to Rostow's drive to maturity and the early stages of high mass consumption. International migration (primarily a transatlantic phenomenon), and frontierward migration (primarily a New World phenomenon), and rural-urban migration all decline as the sources dry up with rural depopulation, or the destinations become less welcoming or attractive. Circulation increases. In this stage population growth rates slow down as mortality rates begin to level off at low levels and fertility rates decline.

*4 The Advanced Society*
This corresponds to the stage of high mass consumption. Resi-

dential migration consisting primarily of inter-urban and intra-urban movement levels off at a high level. Mass international migration ceases with movement confined mainly to people with special qualifications. The volume of circulatory traffic increases at an accelerating rate. Communication becomes very important and increases rapidly. In this stage of the demographic transition population grows slowly or not at all as fertility and mortality rates offset each other at low levels.

## THE GREAT TRANSITION

Perhaps the boldest of all recent attempts to summarize the history of development and project it into the future is 'The Great Transition' of Herman Kahn. In 'The Next 200 Years', published in 1976 to coincide with the bicentenary of the United States, Kahn produced what might be termed a patriocentric model of the world's economic history and future; the model's subtitle is 'A Bicentennial and/or Realistic Perspective of the Prospects of Mankind (in fixed 1975 dollars)'. As a model of the world past and future it has a pleasing symmetry. Before 1776, the year in which economic progress begins, the whole world is described as pre-industrial. By 2176, the model predicts, the whole world will have achieved 'full development'. The process is illustrated in the book by a neat logistic (S-shaped) curve with 1976 precisely at the inflection point.

This is how Kahn summarizes it:

200 years ago almost everywhere human beings were comparatively few, poor and at the mercy of the forces of nature, and 200 years from now, we expect, almost everywhere they will be numerous, rich and in control of the forces of nature. ... We expect that almost all countries eventually will develop the characteristics of super- and post-industrial societies.(10)

## DESIRABLE AND INEVITABLE

Zelinsky contends that the mobility and demographic transitions are both 'thus far irreversible':

Both transitions seem to have a fatalistic inevitability; all human communities have been launched upon them, and if they can surmount the developmental crisis that occurs in midstream, all appear destined to rush forward to whatever terminal conditions may be implied by extremely advanced demographic development.

Similar claims are implicit in the other models. Figure 2.2 is a composite model of the visions of progress embodied in the models of Rostow, Friedmann, Taaffe, Morrill, Gould, Zelinsky and Kahn. All the modellers would attach caveats to their visions, such as Rostow's 'barring global catastrophe', or Zelinsky's 'if they can

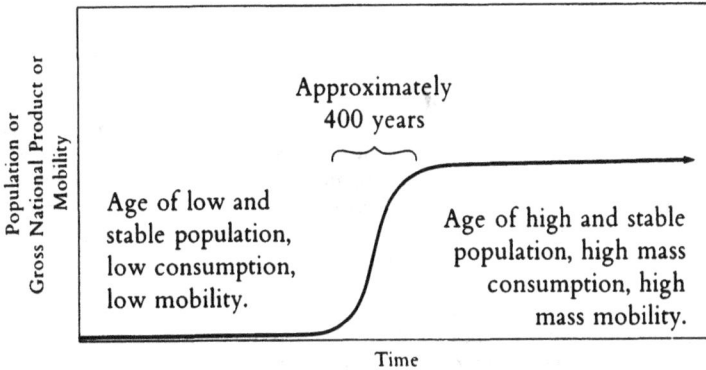

Figure 2.2 Three transitions

surmount'. Some might quibble with Kahn's timetable for future development, and others might argue about the mechanisms of development and the relative importance of the different factors responsible for it, but they are all agreed on the essential attributes of the development process as it has unfolded so far; they all share the view that it is the same ladder of progress that all countries are attempting to climb. They would all agree, with Rostow, that 'it made the life of the average citizen better over the past three centuries, by standards the overwhelming majority of mankind would accept'.

They also all impute a scientific lawfulness to the trends they describe. They all suggest, like Marx, that so long as mankind can refrain from gross violation of the laws of history, the laws will ensure a safe passage to a desired end. They all, like Marx, impute a qualified inevitability to what they consider desirable, and desirability to what they consider inevitable.

In his most recent book, 'The World Economy: History and Prospect', (11), Rostow shows signs of taking his caveat more seriously. In his final chapter he dwells on what he calls the 'Faustian dangers' of technological progress; and he concludes in the manner of one needing some reassurance that 'all societies' are still on course for 'the stage of high mass consumption':

As often in the past, I am driven finally back on these lines of Walt Whitman:
  One thought ever at the fore -
  That in the Divine Ship, the World
    breasting Time and Space,
  All peoples of the globe together sail,
    sail the same voyage.
  Are bound to the same destination.

## A DIFFERENT PERSPECTIVE

It seems unlikely that the destination that Walt Whitman had in mind for the Divine Ship was Rostow's stage of high mass consumption. A secular heaven of high mass consumption is a most improbable destination for a Divine Ship. Nevertheless the models described above have proven useful navigational aids to the past. But they are for a voyage which, thus far, has been experienced by most of humankind as galley slaves, or, at best, passengers in steerage. History viewed from the lower decks looks very different. From this perspective the improvement in living standards of the 'average' passenger charted by the rising graphs of The Great Transitionalists is but a taunting statistical abstraction.

The vision of history embodied in the three transitions of Figure 2.2 can be a comforting one. Transition implies movement from one state to another, and in the midst of rapid unsettling changes, such as those depicted by the graphs of Chapter 1, it is reassuring to be told that one is moving purposefully upward towards a desirable destination, that the curves will all bend over and come gently to rest on the plateau of universal affluence. It is especially reassuring if one is in the vanguard of this movement. It is a vision that can also engender pride, if one is of Western European stock. The idea that global development is a diffusion process that began in Western Europe, that spread out to the rest of the world, and that will ultimately embrace the whole world, has obvious appeal for anyone who can claim some affiliation with the initiators and principal agents of this process.

But, as P. T. Bauer has suggested above, it is a vision that engenders less pride in those on the receiving end of the diffusion process. Diffusion studies are highly selective. What is selected for study appears to be strongly related to the student's perspective on 'the problem'. In Britain, for example, the increase and spread of the country's black population is commonly discussed as a problem. There is considerable popular and governmental concern about it, and this concern is reflected in diffusion studies. In Birmingham, to cite a geographical study of 'the problem', the diffusion of an alien group is discussed in terms of 'residential invasion', 'colour shock' and 'violent stress'.(12) Most liberal and not so liberal solutions to 'the problem' share the objective of minimizing the impact of the alien culture.

The spread of 'progress' into the interior blank areas depicted in Stage 1 of the Ideal-Typical Sequence of Development was accomplished militarily. In West Africa, the area upon which the model was primarily based, it involved among other things the sacking of Kumasi in Ghana and the destruction of a well developed and integrated society centred on Bida in Nigeria. Such events, which could only have been seen as disasters by the Ashanti and the Nupe, are all subsumed by a sequence called ideal and typical. If maps could be drawn for the same parts of the world showing the diffusion of alcoholism, or prostitution and venereal disease, or psychological disorders there is some evidence for believing (13, 14, 15) that they would also depict patterns of the Ideal-Typical variety. Aimé Césaire, one of the most eloquent spokesmen for those on the receiving end of the French variant of the Ideal-Typical sequence, said that when he talked of colonialism he was 'talking of millions of men who have been skillfully injected with fear, inferiority complexes, trepidation, servility, despair and abasement'.(16)

The global developmental diffusion process was led by Britain, and its geographical progress can be estimated from the expansion of its trading hinterland described in Table 1.1. By definition the contact between Europeans and the traditional societies that provided the initial stimulus to change was not a meeting of equals. In the relatively empty parts of the world, such as Australia, South Africa and most of North and South America, the indigenous societies were simply overrun; they were culturally and sometimes physically obliterated. In more populous parts of the world the contact most commonly took the form of colonialism. The economies of the colonies were developed as adjuncts to the economies of the metropolitan powers. The transport and communication systems that were developed to serve these economies were designed to serve the interests of the dominant partner.

Transport and communications were developed explicitly and unashamedly with the primary purpose of exploiting the natural resources of the traditional societies that had been acquired as colonies. They were preconditions for a form of development that was radically different from that of Western Europe.(17) It was a form with a distinctive geographical pattern. The internal transport and communications systems of these countries were designed and built to funnel the raw materials they produced to their port (usually capital) cities. In the international hierarchy of colonialism the colonial economies functioned as lower order appendages to the economies of the colonizers, and the port cities were the points at which the appendages were attached. The legacy of these developments still hangs like a millstone around the development aspirations of these countries.

Individually they constitute markets that can support very little in the way of modern, specialized manufacturing industry. Hence breaking out of their role as suppliers of cheap raw materials to the industrialized world requires the development of trading links among themselves. These are frustrated at every turn by the geographical, economic, cultural and linguistic divisions imposed by the colonial powers, by the jealousy with which these powers continue to guard their economic interests in these areas, and by the colonial transport and communications infrastructures that they have inherited.

Most countries of the 'developing' world have stuck at Stage 2 of the Ideal-Typical sequence. Progress in the industrialized world beyond this stage coincided with, and in no small measure depended on, increasing supplies of cheap labour and materials made accessible by the global 'development' process. In the non-industrialized countries of the world, affluence has been achieved only by a tiny elite concentrated in the principal cities. Life for the rest of their inhabitants remains precarious. In 1978, the Food and Agricultural Organization of the United Nations reported the results from 161 countries of its 'World Food Survey'. It found no evidence of a closing of the gap between rich and poor. Among the rich and industrialized countries it found a steadily rising incidence of diseases associated with obesity. Among the poor countries it found evidence that the average consumption of calories had declined since 1974. The World Bank's 'World Development Report 1978' concludes that in the past ten years food production in the world's thirty-four poorest countries has not kept up with population growth and that one-third of the population of the non-industrialized world lives in conditions of 'absolute poverty'. The poorest thirty-four countries are agricultural countries with a growing dependence on imported food, which by 1976 accounted for 21 per cent of their total import bill.

The international expansion of the economic hinterlands of the countries of the industrialized world has created an international urban hierarchy. The parasitic nature of the principal cities of the non-industrialized countries reflects the pressures and temptations placed upon them by their position in this hierarchy. As national capitals they strive to keep up international appearances. They support diplomatic corps, armies, police forces, legislatures, supreme courts, customs and immigration departments, health services, economic planning departments, museums, universities – the panoply of the modern nation state. In doing so they seek to emulate the industrialized nations, often with the advice of experts from these nations. They do so with resources drained from the rest of the country, or with money borrowed from wealthier nations.

The result of these pressures is a flood of migrants from rural areas. Drawn by the conspicuous consumption of the few, the many eke out an existence in squatter settlements on the geographical and economic margins of the capital cities. But, alarming though the volume of this migration is, it is not sufficient to relieve the increasing population pressure and the grinding poverty of the rural areas in which it originates. The preference demonstrated by these migrants for urban unemployment over rural underemployment suggests that the gulf in material standards of living between urban and rural areas is continuing to grow. Development assistance in the form of loans from wealthy countries has resulted in a burden of debt that for the thirty-four poorest countries now amounts to over 20 per cent of their collective gross national products.(18)

These countries' share of total international trade, the principal agent by which development is diffused according to the transition theories of development, has been declining for at least twenty years.(19)

The mobility transition has both encouraged and facilitated a dramatic rearrangement of population and wealth throughout the world. Rural-urban migration, and greatly increased long-distance traffic in people, goods and information, have resulted in a great concentration of population, wealth, and political power in the world's great cities. Opinion about whether this transformation is desirable is sharply divided. Those who are wealthiest and most powerful seem the most inclined to the view that these trends represent progress. The poorest and weakest remain to be convinced.

## REFERENCES AND NOTES

1   Rostow, W. W. (1971 ed.), 'The Stages of Economic Growth; A Non-Communist Manifesto', Cambridge University Press.
2   Ibid., p. 4.
3   Quoted in Myrdal, G. (1968), 'Asian Drama: An Inquiry into the Poverty of Nations', Clinton, Mass.
4   Bauer, P. T. (1969a), The Economics of Resentment, 'Journal of Contemporary History', 4, pp. 53-4.
5   Bauer, P. T. (1969b), Million-World Pamphlet, 'Spectator', 10 January, pp. 44-5.
6   Bauer, P. T. (1969), op. cit., p. 60.
7   Taaffe, E. J., Morrill, R. L. and Gould, P. R. (1963), Transport Expansion in Underdeveloped Countries: a Comparative Analysis, 'Geographical Review', 53, pp. 503-29.
8   Friedmann, J. (1966), 'Regional Development Policy', Cambridge, Mass.
9   Zelinsky, W. (1971), The Hypothesis of the Mobility Transition, 'Geographical Review', vol. 61, no. 2, p. 221.
10  Kahn, H. (1976), 'The Next 200 Years', William Morrow & Co., p. 1.
11  Rostow, W. W. (1978), 'The World Economy: History and Prospect', Macmillan. Iran is singled out in the book as 'one of the success stories of the developing world'.
12  Jones, P. N. (1970), Some Aspects of the Changing Distribution of Coloured Immigrants in Birmingham, 1961-1966, 'Transactions of the Institute of British Geographers', 50, pp. 199-219.
13  Dumont, R. (1968), 'False Start in Africa', Bungay (first ed. in French 1962), p. 41.
14  Caldwell, J. G. (1969), 'African Rural-Urban Migration', London, p. 107.
15  Fanon, F. (1967), 'Black Skin White Masks', New York (first ed. 1952), and 'The Wretched of the Earth', Harmondsworth (first ed. 1961).
16  Césaire, Aimé, 'Discours sur le Colonialisme', quoted in translation in Fanon (1967), op. cit., p. 9.
17  For an impressively comprehensive survey of these differences see Gunnar Myrdal's 'Asian Drama', Penguin, 1968.
18  World Bank (1978), 'World Development Report', Table 1.
19  Ibid., Table 6.

# 3  CITIES I:
## transport and the
## urban transition

The final mission of the city is to further man's conscious
participation in the cosmic and the historic process. Through
its own complex and enduring structure, the city vastly aug-
ments man's ability to interpret these processes and take an
active, formative part in them, so that every phase of the
drama it stages shall have, to the highest degree possible,
the illumination of consciousness, the stamp of purpose, the
colour of love. That magnification of all the dimensions of
life, through emotional communion, rational communication,
technological mastery, and above all, dramatic representation,
has been the supreme office of the city in history. And it
remains the chief reason for the city's continued existence.

Lewis Mumford (1)

## THE URBAN TRANSITION

To the vision of the cosmic and historic process embodied in the
transitions discussed in the previous chapter, can be added the
urban transition. This transition can be comfortably accommodated
by the same S-shaped graph of progress (Figure 3.1).

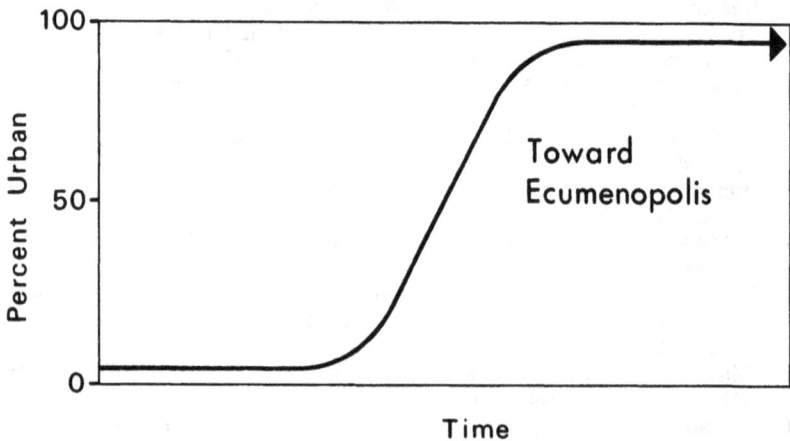

Figure 3.1 The urban transition

Economic productivity above the very low levels found in traditional societies requires the co-ordination and geographical concentration of specialized economic activities. Because towns and
cities are the places where this takes place, an increase in the
proportion of a country's population living in urban areas is a
necessary concomitant of economic development. An acceleration
in this increase is considered by Rostow as one of the principal
signals that the stage of take-off has been reached. Thus urbanization is very commonly employed as a barometer of economic
progress. However, it is an extremely unreliable barometer.

In order to measure urbanization one must have a quantifiable
definition of it, and this, almost invariably, is in terms of the
percentage of a population living in settlements having more than
some minimum number of inhabitants. The minimum population
standard employed to define urbanization can vary; 5000, 10 000
and 25 000 are commonly used standards to distinguish urban
places from non-urban places. But if the population of a country
is growing rapidly in both rural and urban areas, such a definition cannot distinguish an increase in the spatial concentration
of the population from mere population growth. In many parts of
the world, especially the poorest parts, high rates of urbanization
measured in the conventional way engender false hopes. They
simply reflect high population growth in rural areas converting
swollen agricultural villages into 'urban areas'.(2)

The earth-shrinking advances in transport and communications
described in Figures 1.1 and 1.2, which in the early stages of the
urban transition promote a geographical concentration of population in towns and cities, towards the latter stage promote a
dispersal. Cheaper and faster transport for people and goods,
and enhanced abilities to exchange and process information permit
a continued increase in specialization in production, and further
centralization of the control of economic activity, while at the same
time residential populations disperse outward from the focal points
of this specialization and control. And with the growing importance of electronic data processing and exchange, the focal pattern
of specialization and control for some forms of economic activity is
ceasing to have any clear geographical expression at all. Where
the commuting, computing and communicating hinterlands of cities
spread beyond official urban boundaries their residential populations, which are functionally urban, tend to be classified as
rural and the conventional measure of urbanization again becomes
inadequate. Thus the conventional measure tends to overestimate
the rate of urbanization of the poorest countries and underestimate it for the richest.

The end state of the urban transition has been called Ecumenopolis: 'a universal city which will consist of the entire inhabitable
area of the earth'. This is how it is described in 'Spatial Organization' by Ronald Abler, John Adams (no relation) and Peter
Gould (co-author of the Ideal-Typical sequence of Development):

This world city will develop over the next fifty to seventy-
five years, and will be one interconnected functionally

continuous (though not spatially contiguous) urban system.
Time- and cost-space convergence will continue until the
space of the earth has shrunk to the extent that all people
will be as accessible to each other as are contemporary
residents of a metropolitan area. Improved movement systems
will make it possible to construct this worldwide network of
urban life.(3)

Although in its timing this vision of the ultimate stage of the
urban transition is between 125 and 150 years ahead of the final
stage of Kahn's Great Transition, it nevertheless fills in some of
the geographical detail missing from the abstractions of the other
transitions. Although problems are acknowledged, the same tone
of predestination humanly assisted creeps into its presentation.
The highly mobile, high-consuming masses of the advanced society *will* be residents of Ecumenopolis. And this is what it *will be*
like:

For purposes of sending and receiving information, all places
in the world will be located at a single point, just as telephone subscribers within a metropolitan area are now located
at a single point in communications space - that is to say, the
cost and speed of reaching anyone within the city are the
same. We will create the universal city by producing an
isotropic communications and information surface over the
entire earth.(4)

Constantinos Doxiadis, the initial popularizer of the vision of
Ecumenopolis, is even less restrained. He writes as a crusader on
the side of history. The questions of whether the trends described in The Great Transition can, should and will continue are
all muddled together and answered with a single loud 'Yes'.

We are already on the march towards the universal city.

To counteract the trends leading towards the Ecumenopolis
would be both undesirable and impracticable ... the great
forces shaping the Ecumenopolis, such as economic, commercial, social, political, technological and cultural are
already being deployed and it is too late to reverse them.

He even conscripts ignorant faith to march with his army: 'these
are trends of population growth determined by many biological
and social forces which we do not even understand let alone dare
countermand'.

The work of Doxiadis deserves scrutiny because he is much more
specific than the other transitionalists about what the world will
look like when the transitions have run their course. He follows
the logic of the transitions through to its ultimate geographical
conclusion:

We must understand that we will need completely new networks of transportation in which the maximum speeds will be
of the order of hundreds of kilometres or miles per hour. If
we turn to the circulatory system of mammals we find that
while the speed in the capillaries is 0.5-1 mm/sec., in the
veins it is 20 cm/sec., in the carotids 33 cm/sec., and in the
aorta 44 cm/sec., or about 700 times more than in the cap-

illaries. Could not the Anthropocosmos which has speeds of
5 kilometres per hour in its capillaries reach this level of
organization and come to have central aortas with speeds of
3500 kilometres per hour?
The Ecumenopolis will be hierarchical in structure. It will
probably have one centre of administration for all international affairs. However it will certainly contain many others
of great international importance as well as many completely
specialized centres for research, the care of health, philosophy etc. It will then have centres of a different order, from
large to very small, serving people in their various neighbourhoods. It will have facilities of different orders, from the
ecumenic highways and networks of communications down to
the small residential roads and parks, from large natural
parks, down to small local parks at short distances from
every house.(5)


PROGRESS SO FAR

In 'Great Cities and Their Traffic', first published in 1977,
Michael Thomson presents the findings of the most comprehensive
survey ever undertaken of urban transport problems. It contains
a wealth of evidence from all around the world for judging the
progress so far of the urban and mobility transitions.
  Thomson approached his materials with the conviction that sound
planning could permit civilized cities to coexist with mass car
ownership. In Halfway to a Motorized Society he wrote, in 1971,
  New roads; railways, airports, shopping centres, new towns
  and other major building developments can and must be
  planned on the assumption of general car ownership.
  The main advantages of the car can be enjoyed and the
  worst disadvantages avoided by policies designed to resist
  the undesirable trends inherent in motorized societies -
  notably those towards dispersal and the run down of public
  transport - and by policies of traffic restraint and traffic
  avoidance in large towns and cities.(6)
In 1977, after touring the great cities of the world, he had virtually no successes to report:
  The universal similarity of the problem is remarkable when
  one considers that some cities are much richer than others,
  some much older than others, some have extensive rail systems while others have none, some have enormous freeway
  networks while others have built no freeways at all. Yet,
  with few exceptions, all suffer from severe traffic congestion,
  parking difficulties, ailing public transport systems that are
  overcrowded during the rush hours and in financial difficulties, high accident rates, unsatisfactory conditions for
  pedestrians, and degredation of the environment through
  noise, pollution danger and ugliness.(7)
Wilfred Owen, in 'The Accessible City', arrives at a similar

conclusion: 'In all the world's major cities, from Bogota to Bangkok to Boston, the conflict between the city and the car is at the point of impending crisis.'(8)

Everywhere in the world urban traffic problems are bad, and everywhere there is a conviction among transport planners that they will get worse. The reason for this pessimism is, paradoxically, an optimism about the world's economic prospects: 'All cities are expecting continued growth in personal income and hence in car ownership rates. To some extent the level of car ownership is subject to influence by government policy, but most governments regard the general desire for car ownership as irresistible.'(9)

Despite the 'universal similarity' of their traffic problems the cities of the world also have some significant differences. Thomson, with the use of five urban transport archetypes, brings a useful order out of the cacophony of the world's traffic jams. His archetypes, with the cities he has chosen to represent them, are as follows:

Full motorization – Los Angeles, Detroit, Denver and Salt Lake City;

Weak centre – Melbourne, Copenhagen, San Francisco, Chicago and Boston;

Strong centre – Paris, Tokyo, New York, Athens, Toronto, Sydney and Hamburg;

Traffic limitation – London, Stockholm, Vienna, Bremen, Goteborg, Singapore and Hong Kong;

Low cost – Bogota, Lagos, Calcutta, Istanbul, Karachi, Manila and Tehran.

These groups of cities can be assigned in descending order to positions on the ladder of car ownership. At the top are found the cities more or less completely committed to a car-based transport system. At the bottom are the cities in which, despite a few impressive but highly localized traffic jams, the principal mode of transport for most people is still walking. About halfway between these two groups in terms of car ownership levels are the traffic limitation cities in which public and private transport coexist in a state of mutual antipathy. The cities in the remaining two groups towards the top of the order have transport systems and transport policies that reflect varying degrees of unstable compromise with the commitment to full motorization.

## THE TOP

The representatives of the fully motorized archetype chosen by Thomson are all American. It is difficult to over-emphasize the uniqueness of the United States's experience with the car. At the turn of the century there were only about 8000 cars in the whole country; they were more technological curiosities than a form of transport. By 1975 there were 106 million cars in the United States; not only did the United States have exceptional amounts

of land for people to move into and build upon, during this period it totally dominated the world's automotive fuel supply industry. As recently as the early 1970s the United States was at the
same time the world's leading producer of oil and its leading
importer, and the production and marketing of the imported oil
was the almost exclusive preserve of American companies. In the
first three-quarters of the century the population almost trebled,
increasing from 75 million to 215 million. In spite of this rapid
population growth the United States still has, by world standards,
a low population density, at fifty-eight per square mile approximately one-tenth the population density of the United Kingdom.
The first three-quarters of the century were also a period
of rapid urbanization and westward migration. The United
States grew up with the car, and was built up for the car, in
circumstances that no other country in the world can hope to
replicate.

*Table 3.1  Urban growth rates and levels of car ownership*

| City | Population 1900 | Population 1970* | Growth X | Cars per capita 1970 |
|---|---|---|---|---|
| Los Angeles | 102 479 | 7 041 980 | 68 | .48 |
| Detroit | 285 704 | 4 435 051 | 16 | .48 |
| Salt Lake City | 53 531 | 705 458 | 13 | .42 |
| Denver | 133 859 | 1 239 477 | 9 | .46 |
| San Francisco | 342 782 | 3 108 782 | 9 | .43 |
| Boston | 560 892 | 3 848 593 | 7 | .33 |
| Chicago | 1 698 575 | 6 977 611 | 4 | .32 |
| New York | 3 437 202 | 9 973 716 | 3 | .29 |

\* Standard Metropolitan Statistical Area

Sources: 'World Almanac 1976', and 'Transportation Planning Data for Urbanized Areas',
US Department of Transportation (1973).

Table 3.1 illustrates some of the salient features of this growth,
and its consequences in terms of car ownership, for the American
cities chosen by Thomson to serve as archetypes. Generally, the
cities that have grown fastest in this century are those farthest
to west; they are also the most dependent for their transport on
the private car. Car ownership began with the very wealthy and
then spread to the middle classes. The enormous physical expansion of American cities in this period was also led by the better
off. They typically built new homes for themselves while the less
well-off took possession of their residential cast-offs, and occupied them at much greater densities. Thus in the moulding of the
new residential land-use patterns of American cities, car owners
exerted a disproportionate influence.

*Los Angeles*
Los Angeles is manifestly the most attractive major city in the
United States, judged by the rate at which it has attracted new
residents. It is now second in size to New York and still gaining
rapidly. This demonstration of the pulling power of Los Angeles
is considered by many to be conclusive evidence that the city is
providing what people want, and that it ought, therefore, to
serve as an example for the rest of the world. This argument is
exemplified by the following passages from Christopher Rand's
'Los Angeles: The Ultimate City' (Oxford University Press, 1967):
> The market place and the government are different ways of
> serving the only end a democracy can serve - providing what
> people want. Sometimes the best way to do this is through
> business enterprise and free markets. Sometimes, govern-
> ment is the best way. Sometimes people vote with their
> dollars; sometimes with ballots. But the goal is always the
> same - providing what people want. As far as urban trans-
> portation is concerned what people want is clear. They have
> voted overwhelmingly in favour of the automobile.

> To be successful urban transportation planning must be in
> harmony with the overwhelming mandate the people have given
> the automobile. The purpose of metropolitan planning is not
> to provide what the planners decide people should have -
> even if they don't want it - but to develop better ways of
> giving people what they do want and are willing to pay for.

> The fact is ... that for most travel purposes, no vehicles
> have yet been developed or are even in prospect that equal
> the automobile for speed, comfort, convenience, privacy,
> economy and other qualities that people value.

This view, that the land-use pattern of the United States is the
product of a motorized democracy, is also found in John Rae's
'The Road and the Car in American Life' (MIT Press, 1971). He
defends suburbia against its critics and argues that the democratic verdict of the referendum on alternative living environments
has been won decisively by the suburbs: 'American urbanites who
are in a position to do so vote decisively for suburbs with their
cars.'
Another observer who stresses the democratizing powers of full
motorization is Reyner Banham, author of 'Los Angeles: The
Architecture of Four Ecologies'. This is how he puts the case:
'[the] private car and the public freeway together provide an
ideal ... version of democratic urban transportation: door to door
movement, on demand, at high average speeds over a very large
area.'
Banham is exhilarated by the possibilities offered by Los
Angeles, especially for architects; 'pedestrian' is his epithet for
people who do not like it:
> There is ... still a strong sense of having room to manoeuvre.
> ... Los Angeles has room to swing the proverbial cat, flatten
> a few card-houses in the process, and clear the ground for

improvements that the conventional type of metropolis can no longer contemplate.

This sense of possibilities still ahead is part of the basic life style of Los Angeles. It is, I suspect, what still brings so many creative talents to this palm-girt littoral - and keeps them there. For every pedestrian litterateur who finds the place 'a stinking sewer' and stays only long enough to collect the material for a hate-novel, for every visiting academic who never stirs out of his bolt-hole in Westwood and comes back to tell us how the freeways divide communities because he has never experienced how they unite individuals of common interest ... for these two there will be half a dozen architects, artists or designers, photographers or musicians who decided to stay because it is still possible for them to do their thing with the support of like-minded characters and the resources of a highly diversified body of skills and technologies. (10)

The vision of a city consisting of communities without geographical expression, united by common interests, had been most fully elaborated by Melvin Webber in an article entitled Order in Diversity: Community Without Propinquity. He characterizes the critics of the motorized city as people in the grip of 'obsolete truths'. His vision of future society is one in which the old-fashioned notion of community, as a physical place in which people of all ages and interests live and know each other, is replaced by the concept of spatially dispersed 'communities of interest'. People, freed by advances in transport and communications from their burdensome relationships with incompatible geographical neighbours, will live their lives in the almost exclusive company of members of their particular affinity group.

This emerging community is to be found in its most advanced form in southern California:

As the urban freeway systems that are now under construction are extended farther out and connected to one another, an unprecedented degree of freedom and flexibility will be open to the traveller for moving among widely separated establishments in conducting his affairs. A network of freeways, such as that planned for the Los Angeles area, will make many points highly accessible.... And the positive advantages of automobiles over transit systems ... make it inconceivable that they will be abandoned for a great part of intrametropolitan travel or that the expansion of the freeway systems on which they depend will taper off. We would do well, then, to accept the private vehicle as an indispensable medium of metropolitan interaction - more, as an important instrument of personal freedom. (11)

An alternative view of the end result is provided by Michael Thomson:

Apart from the hilly suburbs of Hollywood in the north, much of Los Angeles is flat and laid out in grid patterns. Excepting the downtown area, the city is mostly low-level

and dreary. The environment is often ugly and depressing,
noisy and dirty - mile after mile of drab streets in rect-
angular blocks, festooned with hoardings, telegraph posts,
traffic signs and trash - lifeless streets without a soul on them
except those boxed away in cars.... During the evening only
about 40,000 or about 0.05 per cent of the population, were
there [in the city centre], of whom a not inconsiderable proportion can be clearly identified as social dropouts of one
kind or another. (There are an estimated 1.4 million alco-
holics in Los Angeles and a similar number of drug addicts.)
Visitors are advised not to venture alone on the streets after
8 pm. It can truly be said therefore that downtown is of no
consequence to the great majority of Los Angelenos. It stands
out as a centre only because there are no other sizeable
centres in the metropolis. There are few recognizable communities. It is for this reason that one can drive for tens of
miles in Los Angeles without ever seeming to get anywhere.(12)
Another critic of Los Angeles is Lewis Mumford. It is, in his
view, the antithesis of all the things a city ought to be:
In the mass movement into suburban areas a new kind of community was produced, which caricatured both the historic
city and the archetypal suburban refuge: a multitude of
uniform, unidentifiable houses, lined up inflexibly, at uniform distances, on uniform roads, in a treeless communal
waste, inhabited by people of the same class, the same
income, the same age group, witnessing the same television
performances, eating the same tasteless pre-fabricated foods,
from the same freezers, conforming in every outward and
inward respect to a common mold, manufactured in the central
metropolis. Thus the ultimate effect of the suburban escape
in our time is, ironically, a low-grade uniform environment
from which escape is impossible.
   The absurd belief that space and rapid locomotion are the
chief ingredients of a good life has been fostered by the
agents of mass suburbia.... The reductio ad absurdum of
this myth is, notoriously, Los Angeles.(13)

*Detroit*
Los Angeles's principal competitor as the exemplar of the motorized city is Detroit. Although it contains fewer inhabitants than
Los Angeles it earns its popular title of 'Motor City' twice over.
As Table 3.1 shows, on a per capita basis it owns approximately
as many cars as Los Angeles; it is also the world centre of the
car industry. On display in Detroit in exaggerated form therefore, are the consequences not only of extremely high levels of
car use, but of extreme dependence on the pre-eminent economic
activity of a motorized society.
   Popular opinion about American cities such as Detroit tends to
be highly volatile. In 1965 the American magazine 'Time' was of
the opinion that its problems were being tackled 'with vigor and
imagination'. This is how it viewed Detroit in the issue of 24

September 1965: 'The heart of the city, half dead in 1961, is pulsing again with new office buildings and hotels and the return of many suburbanites to new luxury apartments.'

Less than two years later, in the issue of 4 August 1967, its view was less sanguine:

In the violent summer of 1967 Detroit became the scene of the bloodiest uprising in half a century and the costliest in terms of property damage in U.S. history. At the week's end, there were 41 known dead, 347 injured, 3800 arrested. Some 5000 people were homeless ... while 1300 buildings had been reduced to mounds of ashes and bricks, and 2700 businesses sacked. Damage estimates reached $500 million.

In the early 1960s the work of Detroit's transport planners was frequently held up to Europeans as an example to emulate; in 'London 2000' it was argued that 'The Detroit example should provide a model for a "first stage" programme of motorways for London'.(14) Its history of development is now more commonly consulted for the insights it might provide into mistakes which, it is fervently hoped, might be avoided.

The United States achieved a level of car ownership in 1920 that was not reached in Britain until 1955, about 7.5 cars per 100 of the population. The late 1940s and early 1950s were the years in which most public transport facilities in British cities reached their all time peak in terms of the provision and use of public transport, since which time they have experienced continuous decline. Table 3.2, taken from 'A History of Detroit Street Railways' published in 1931, illustrates how rising car ownership made an impact on urban public transport systems in the United States very similar to that which occurred three decades later in Britain.

The lamenting in 1931 about the impact on the Detroit public transport system of increasing numbers of cars has a remarkably modern ring in British ears:

The Detroit street railway in 1930 is the largest system in the world under public ownership.... The Department of Street Railways has been operating the property under the most adverse conditions. Competition such as was never dreamed of in the old days exists in the form of privately owned and operated motor vehicles, tens of thousands of them, carrying not only the owners, but thousands of fellow workers and friends of the owners, who otherwise would be patrons of the street railway. Congestion of sur-face traffic in down-town streets has brought about extremely difficult operating conditions and has resulted in a system of traffic regulation by means of lights which has not only slowed the movement of street cars, but which makes it impossible to provide fast schedules on any lines of the system on sections of track several miles in extent.(15)

The Detroit street railway system, whose traffic at its peak amounted to almost one journey a day for every one of its citizens, was truly a 'mass' transport system. It now no longer exists. The

Table 3.2 Comparative operating statistics by years (revenue passengers per 1000 population)

| Year | Kansas City | Minn. St Paul | Baltimore | St Louis | Detroit | Cleveland | Cincinnati |
|---|---|---|---|---|---|---|---|
| 1920 | 289 017 | 265 698 | 333 422 | 352 986 | 333 724 | 360 892 | 238 243 |
| 1921 | 283 715 | 336 317 | 307 890 | 340 707 | ...... | 306 055 | 214 276 |
| 1922 | 283 886 | 325 276 | 297 748 | 340 906 | ...... | 296 489 | 215 447 |
| 1923 | 271 280 | 308 405 | 299 559 | 344 593 | 288 360 | 298 182 | 216 784 |
| 1924 | 248 371 | 283 646 | 289 812 | 324 003 | 273 655 | 269 416 | 200 757 |
| 1925 | 232 760 | 271 329 | 272 793 | 305 604 | 248 075 | 258 781 | 193 104 |
| 1926 | 215 216 | 247 616 | 271 720 | 297 558 | 250 765 | 244 724 | 190 007 |
| 1927 | 203 309 | 228 799 | 259 140 | 278 907 | 208 500 | 225 500 | 199 166 |
| 1928 | 196 241 | 215 771 | 234 464 | 262 015 | 185 785 | 216 982 | 200 492 |

Source: O'Geran, G. (1931) 'A History of the Detroit Street Railways', Conover Press, Detroit.

people who are dependent on the bus service, which is all that remains of public transport in Detroit, suffer a service that is dirty, dangerous, and unreliable, and which leaves large areas of the city virtually inaccessible. They are the people who were left behind in the flight to the suburbs. Because the decline of the public transport system in Detroit began so much earlier than in any other country in the world, there is now no way in which it could be revived. Such a large proportion of the city's built environment has been designed to accommodate the car, that most of its population is now too dispersed to be adequately or efficiently served by any conceivable system of public transport.

Detroit has a history of violence. It is a history of extremes of affluence and poverty and the tensions and conflict they have created. In 'American Earthquake' Edmund Wilson wrote of Detroit in the 1930s:

> The huge organism of Detroit, for all its Middle Western vigor, is clogged with dead tissue now. You can see here, as it is impossible to do in a more varied and complex city, the whole structure of an industrial society; almost everybody who lives in Detroit is dependent on the motor industry and, in more or less obvious relation, to everybody else who lives here. When the industry is crippled everybody is hit.(16)

Wilson describes the way in which a mass production industry consisting of expensive machines tended by relatively unskilled and poorly paid workers produced Detroit's extremes of affluence and misery. When business was good huge profits accrued, largely to the owners of the machines. When business was bad, and it required only a slowing down in the growth rate of the car population for it to turn bad, the machines were turned off and the workers were turned out. The extremely uneven distribution of wealth that resulted acquired a clear geographical expression. Those with the ability to do so, turned their backs on the problems of the inner city and took their houses and their factories to the suburbs:

> As for Henry Ford himself, his reputation as a benefactor of the American working man has conspicuously declined. His removal of his factories to Dearborn outside the city limits, in order to escape city taxes, has relieved him from contributing anything to the relief of the unemployed, a third of whom, according to the city's calculations, have been laid off from his own plant.(17)

In the aftermath of the 1967 riots Detroit's civic leaders and captains of industry, in what they hoped would be an inspirational gesture, built the Renaissance Centre, an impressively glossy hotel-office-shopping complex, in downtown Detroit within the hub of the city's radial motorway network. But Henry Ford's legacy remains intact. It was a gesture that simply reaffirmed the enormous economic distance which separates the city's 'haves' from its 'have nots'. The journey along one of Detroit's radial roads such as Kercheval, from the Renaissance Centre to its suburb of Grosse Pointe, is a journey from commercial splendour, through burned-out squalor, to residential opulence.

Detroit is justifiably famous for its nastiness and violence, and this fame assists the impression that there must be something uniquely wrong with the city. But in terms of the official American league table of nastiness and violence, the FBI's 'Union Crime Reports', the city is typical of many others. In 1972, statistics for the 227 Standard Metropolitan Statistical Areas showed that per capita it ranked only fifteenth in terms of car theft, twenty-second in terms of murder, and thirty-eight in terms of rape.(18)

Detroit's riots were merely the most expensive and destructive of more than 150 'civil disorders' that took place in the United States in the summer of 1967. On 11 August 1967, a week after the report quoted above, 'Time' provided the following survey of the American urban scene:

Detroit was a burned out volcano, and although Milwaukee trembled, its authorities hammered down an iron lid that saved the city from massive hurt. Still there was little peace in the nation's cities. From Providence R.I. to Portland Ore., communities large and small heard the sniper's staccato song, smelled the fire bomber's success, watched menacing crowds on the brink of becoming mindless mobs.

Although there has been no rioting on the scale of the 1967 disorders since that time, criminal violence is endemic in American cities. It is increasingly accepted as a fact of life. The wave of arson and looting that swept through Brooklyn, Harlem and the Bronx during the power failure in July 1977, suggests that large-scale disorder threatens whenever the iron lid hammered down by the authorities comes loose.

Edmund Wilson suggests that Detroit is a model city; it represents in simplified and exaggerated form the ills of the larger urban industrial society of which it is but a part:

Its lines of development are clear, and its complexities comprehensible. Over the course of a century it was involved in many major historical developments. As the world centre of the automobile industry, it has exerted tremendous social and economic influence. It has been headquarters of two of the world's three largest corporations, and it is America's most unionized city. Detroit is the heart of American industry, and by the beat of that heart much of America's economic health is measured.(19)

There remains fierce disagreement about the moral of the tale told by the fully motorized cities. Some clearly view its ills as but the pains of the late transition, what Herman Kahn has referred to as 'the growing pains of success', which must be endured in order to reach the promised land of Ecumenopolis. Others view the same ills as the products of a pathological process of 'development' which will culminate in grave disorder. The debate has been briefly aired here, not resolved. It is unlikely to be resolved in the near future; the view of the motorized city obtained by one of Reyner Banham's cat swinging creative talents in Los Angeles, or by the inhabitants of Detroit's affluent suburbs, is too starkly different from the view of the same cities from Watts or the downtown end of Kercheval.

But the lesson of these cities for cities in other parts of the
world towards the bottom of the motorization scale is clearer. The
American experience, desirable or undesirable, is not repeatable,
because the conditions in which it occurred are not repeatable.
The cities of the world that exist now and which were built before
significant numbers of their inhabitants owned cars, have, with
few exceptions, neither the undeveloped land, nor the fuel, nor
the money necessary to follow the American example.


## THE BOTTOM

The major cities of the poor world offer a precautionary lesson in
interpreting the evidence of migration. The fact that Americans
have been migrating in substantial numbers to suburbs, and to
cities like Los Angeles, is frequently adduced as evidence of the
manifest superiority of these destinations as forms of urban life.
Such arguments are rarely used in defence of the major cities of
the poor world, which are also the targets of large numbers of
migrants. The attractive force that impells much migration is akin
to that which draws the moth to the flame. The flame is well
exemplified by Reyner Banham's description of life in Los Angeles,
and the grim reality that is the lot of many, by Thomson's account.
In poor cities the contrast between hope and reality is much
sharper.

*Lagos*
Nigeria, with a per capita income estimated in the World Bank's
1978 World Development Report at £190, is classified by the World
Bank as a 'middle income country'. Of the 'low cost' cities surveyed by Thomson, Lagos, in terms of income, lies towards the
middle of the range. It is much wealthier than the poorest, Calcutta, and considerably poorer than the wealthiest, Tehran. It
displays the extremes of wealth and poverty characteristic of all
such cities.
   The capital city of Nigeria is growing very rapidly. Estimates
of its present size are very approximate guesses because no close
track is kept of the numbers of new immigrants. Current estimates
put its population at about 3 million. This is Michael Thomson's
description of it:
   With this huge growth of population, for which the supply of
   housing was grossly inadequate by any standards, most
   services had come near to breakdown. Worst of all was the
   disposal of sewage and refuse. The large open storm drains
   built under British rule were almost totally blocked and over-
   flowing in the streets owing to lack of maintenance. The
   refuse lay in mountains in the middle of residential streets,
   scratched at by scraggy hens. With an average house occu-
   pancy of ten persons per room, thousands of people were
   virtually homeless and slept in the streets amidst the filth.
   The water supply frequently dried up.... The people,

however, were generally well fed and clothed, by African standards.(20)

Both transport and communications in Lagos are in a state of chronic chaos. Traffic congestion on Lagos Island, the focus of the city, is described by Thomson as ranking 'among the worst in the world'. But the car population of Lagos at that time was only 55 000. The amazing congestion of Lagos was being created by a number of cars that would cause scarcely a ripple if added to the traffic of one of Los Angeles's major freeways. Most of them were attempting to converge by five bridges on to a densely packed island with an area of less than 2 square miles (518 hectares), and most of whose streets were laid out for pedestrians. Put another way, Lagos's traffic problem consisted of 50 per cent of the cars in Nigeria, owned by .09 per cent of the population, trying to crowd into .0006 per cent of the area.

While a few tens of thousands of cars were causing near paralysis of traffic in the city, public transport, which was a monopoly of the Lagos City Transport Service, consisted of a fleet of only 250 buses. Because of congestion these buses managed to travel only an estimated 96 kilometres (60 miles) each per day.

The use of bicycles is extremely dangerous. The danger to cyclists from cars, and the inhibiting effect this has on the use of bicycles, is a phenomenon that has been observed in cities almost everywhere in the world, whether they have high levels of car ownership or not. In cities with relatively few cars, drivers make up for their lack of numbers by driving very much more destructively. The Transport and Road Research Laboratory in a report entitled 'A Study of Road Accidents in Selected Urban Areas in Developing Countries', records that in Bombay in 1970, 120 000 vehicles managed to kill 645 people. In 1975 in London with 2.3 million vehicles only 638 people were killed in road accidents.(21) The Bombay kill rate transposed to Britain's present vehicle population would yield about 100 000 killed every year. If the whole of India had American levels of car ownership, and Bombay's fatal accident rate, about 2 million would be killed every year.

There are many explanations proffered for the high death rates per vehicle in poor countries which are discussed in Chapter 7, but perhaps the most obvious cause that occurs to anyone with experience of traffic in such countries, is the sheer arrogant disregard that so many drivers have for the chickens, goats, cyclists and 'peasants' with whom they share the road.

Car ownership patterns in these cities provide a useful index of the concentration of wealth and political power. And official transport plans provide a useful guide to the planning philosophies of those who wield power. Because transport plans have a highly specific geographical dimension they reveal more clearly than plans couched in terms of abstract economic sectors whether their authors adhere to the view of 'development' as a process that diffuses outwards and downwards, or whether they see it as a process that involves building from the bottom up.

Clearly there is no prospect of the overwhelming majority of Nigerians now living ever owning a car. There are however considerable transport improvements that could be achieved within the lifetimes of these people. The improvements of tracks for bicycles and carts and the organization of community transport services based on lorries and buses, for example, could transform the lives of millions. In sharp contrast, a diffusionist philosophy of transport development fixes its eye on a higher, more distant goal. It declares the goal of universal car ownership to be politically irresistible and directs its money and planning energies towards the achievement of this goal - however many generations it might take.

In the case of Lagos, the application of the bottom-up philosophy would result in pedestrians, cyclists and buses being given absolute priority in all transport planning. This appears most unlikely to happen. Thomson reports that at the time of writing the Lagos area masterplan was based upon 'the Californian concept of a grid motorway network'.

*Calcutta*
The ability of Western transport planners to disregard the present needs of the many, in pursuit of a remote high-mobility Utopia, is nowhere more impressively demonstrated than in Calcutta, one of the most desperately poor cities in the world. Here, for most people, the problem of traffic congestion consists of the impediment to the movement of pedestrians and human-pulled rickshaws and carts, created by other pedestrians, rickshaws and carts, and by starving beggars living in the streets. Here is Thomson's despairing summary of transport planning in that city:

> the 1967 plan [included] proposals for a ten-year motorway programme involving 80 kilometres of expressway, two new bridges, and 32 kilometres of major road widening, together with an elevated rapid transit system, electrification of all suburban railway lines, and widespread traffic engineering schemes. The cost of this programme was optimistically estimated at 1,400 million rupees (£70 million), but no consideration was given to the cost of maintaining the expressways and rehousing the thousands of people displaced by them. More important, no thought was given to the operating costs of a rapid transit system; if people cannot afford the cost of an old, overloaded bus, how can they afford the cost of a rapid transit system?(22)

From his survey of poor cities Thomson concluded that cheap and comprehensive bus services offered the only hope of bringing some sort of order out of the transport chaos of these cities. But nowhere could he find evidence of effective support for such an unglamorous solution: 'Unfortunately, none of the cities studied ... has yet shown any sign of recognizing that, for them, this is the only solution.'(23)

## REFERENCES AND NOTES

1   Mumford, L. (1961), 'The City in History', Secker & Warburg.
2   Adams, J. G. U. (1970), 'The Spatial Structure of the Economy of West Africa', Ph.D thesis, University of London, pp. 129-49.
3   Abler, R., Adams, J. and Gould, P. (1972), 'Spatial Organization', Prentice-Hall, p. 555.
4   Ibid., p. 556.
5   Doxiadis, C. (1968), 'Ekistics: An Introduction to the Science of Human Settlements', Hutchinson & Co., London, pp. 430, 459, 460.
6   Thomson, M. (1971), Halfway to a Motorized Society, 'Lloyds Bank Review', October, no. 102, pp. 16-34.
7   Thomson, M. (1977), 'Great Cities and Their Traffic', Victor Gollancz, p. 318.
8   Owen, W. (1972), 'The Accessible City', Brookings Inst., Washington DC.
9   Thomson (1977), op. cit., p. 88.
10   Banham, R. (1971), 'Los Angeles: The Architecture of Four Ecologies', Harper & Row, pp. 242-3.
11   Webber, M. (1963), Order in Diversity: Community Without Propinquity, in 'Cities and Space: The Future of Urban Land', L. Wingo (ed.), Johns Hopkins Press, pp. 23-54.
12   Thomson (1977), op. cit., p. 101.
13   Mumford (1961), op. cit., pp. 486, 510.
14   Hall, P. (1971), 'London 2000', Faber, caption to photograph 20.
15   O'Geran, G. (1931), 'A History of the Detroit Street Railways', Conover Press, Detroit.
16   Wilson, E. (1975), 'American Earthquake', Octagon Books, New York, p. 232.
17   Ibid.
18   Positions in this table change from year to year. In 1976 Detroit placed rather higher, coming second after San Francisco in the total crime index for a list of fifteen selected cities. In 1976 there was one murder, robbery, rape or assault for every nine inhabitants (Table 277, 'Statistical Abstract of the United States').
19   Wilson (1975), op. cit., p. 232.
20   Thomson (1977), op. cit., p. 234.
21   Greater London Council (1979), 'Transport Policies and Programme, 1979-84'.
22   Thomson (1977), op. cit., p. 250.
23   Ibid., p. 262.

# 4  CITIES II:
## transport in London

Midway between the wealthy, fully motorized cities of the western
United States, and the largely pedestrian cities of the poor world,
are the major cities of the industrialized world whose built environments largely pre-date the age of the car. All such cities are in
the process of attempting, with varying degrees of lack of success, to come to terms with the car. London is one of the largest
of these cities and, with one car for every fourth inhabitant, is
almost exactly halfway to the car ownership levels of fully motorized American cities. Although the history of its transport development is unique in its complex detail, it has much in common
with all those cities whose land-use patterns were not intended
for, and are incapable of accommodating, large numbers of cars.
A second reason for choosing London to represent the middle
ground is that it has been my home for the past thirteen years
and is the city with whose transport system I have had the most
direct experience.

To most of us who travel in London, the city's transport system
is bewilderingly large and complex. If we plan excursions that
depart from our regular routes, we usually need to consult one of
the various London maps or directories to find our way. We commonly find, especially if either end of our journey is close to the
centre of the city, that we must choose from a number of possible
routes and modes of travel. Our choice will be influenced by matters such as whether or not we have a car or a bicycle, the
availability of a place to park, the length and cost of the journey,
our knowledge of traffic conditions along the route, and the
weather.

For the majority of their journeys most people would probably
describe their particular choice of route and mode of travel as the
least of a number of possible evils. Traffic jams, fumes, noise and
accidents afflict those who travel by road. Rail and underground
services are often dirty, unreliable and overcrowded. Traffic also
makes life unpleasant in the residential neighbourhoods and high
streets through which it passes.

To the people experiencing them, these problems have an air of
permanence. Although there are occasional changes - road-works,
new schedules and altered services, new one-way systems and
parking regulations - the great bulk of the physical network that
carries the traffic and the institutions that administer it can seem
solid and enduring. If we retreat, however, to a more detached

perspective and contemplate London's transport over the span of
the nineteenth century and the first three-quarters of the twentieth century the picture becomes one of dramatic change. Looking
ahead to the end of this century the potential for change is
equally impressive. The history of this change is intimately bound
up with the story of the growth of London and the continuous
efforts of the planners, operators and users of transport in the
city to cope with a situation that has never been considered satisfactory. The future of London's transport will doubtless be shaped
by a continuation of these same efforts.

THE PAST

*The Nineteenth Century*
'They know it's handy for a man's work' said one and all,
'and that's the reason why they imposes on a body'.
                          from 'Jacob's Island' by Henry Mayhew (1849)
Jacob's Island, in Bermondsey in south-east London, was visited
by Mayhew in 1849. The filth, overcrowding, hunger and disease
described in his stomach-turning account of what he called 'a
Venice of Drains', where the drains served also as the source of
the area's drinking water, provoked an obvious question: how
could people acquiesce in such an existence? Mayhew put this
question to people living there. He encountered a remarkable
stoicism in the inhabitants and an acute appreciation of the causes
of their suffering. The reason why the landlords of Jacob's Island
were able to impose so cruelly on them was that the area was convenient to the inhabitants' work. While this is obviously less than
a complete explanation for the conditions Mayhew found, it does
serve to illustrate what a powerful force the journey to work has
been in shaping London. Of all the bustling to-ing and fro-ing in
London's streets it was, and remains, the journeys to work that
dominate the pattern of movement. These journeys, in turn, have
been an important consideration in the myriad locational decisions
that have produced London's land-use pattern.
    The influence of the journey to work is strongest where people
are poor and transport is expensive. The observations of Charles
Pearson to the Royal Commission on Metropolitan Termini in 1846
are still relevant: 'A poor man is chained to the spot. He has not
leisure to walk and he has not money to ride a distance from his
work.'(1) Pearson was convinced that cheap transport could play
a vital role in breaking the chains that bound people to slums
such as Jacob's Island. He was one of the most energetic of the
promoters of the Metropolitan Line from Paddington to Moorgate,
the world's first underground railway, and obtained an undertaking from its operators, the Great Western Railway, to carry
workers from a proposed housing estate in the country, about 5
kilometres west of London, to the City and back daily for one
shilling a week.
    Although the railway building of the nineteenth century was

primarily spurred by other - less altruistic - motives, it did help
to relieve the overcrowding of the urban poor. London experienced
a sixfold growth in its population, from 1 110 000 at the beginning
of the nineteenth century to over 8 000 000 by its peak in 1951,
but a more than seventyfold growth in its area during the same
period as it sprawled into the surrounding countryside. At the
same time, the demand for land near the centre for industrial and
commercial uses intensified. Thus, partly as a result of people
being pushed out of the centre and partly as a result of the pull
of the new suburbs, the extreme residential densities at the centre
began to decline. By 1961 when the number of people working in
the City was close to its peak of 500 000 there were only 4771
people actually living there.(2) Pearson's workers' estate, however, did not materialize. It was overwhelmingly the better-off
who took advantage of the opportunities afforded by the railways
to live in the suburbs and work in the central city.

It is important to remember how malleable the terms 'suburb' and
'central city' have had to be. The underground Metropolitan Railway from Moorgate to Paddington, completed in 1865, was built to
link the City with its north-western suburbs. It was also intended
to relieve the severe congestion caused by the rapidly growing
volume of horse-drawn traffic along London's northern perimeter
road. Today, congestion is as bad as ever and both terminals of
the railway as well as the 'perimeter road' (the present Marylebone
and Euston Roads) are part of the modern transport planner's
'central city'. Figure 4.1 describes this suburbanization process
from 1841 to 1961. The process of dispersal has continued to the
present, but after the peak reached on the graph in 1951 it spilled

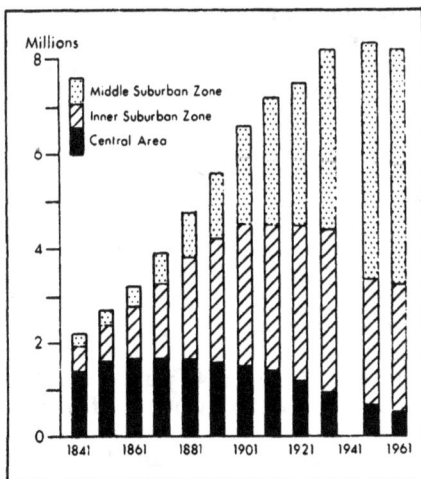

Figure 4.1 London's population growth, 1841-1961
Source: White, H. P. (1971), 'London Railway History', David &
        Charles.

over into the outer metropolitan area so that population statistics
for the GLC area alone are no longer sufficient to chart its progress.

The transformation of London's transport system in the nineteenth century was profound. Until London's first railway - from
Spa Road to Deptford in south-east London - was opened in 1836
all travel over land in the city had to be on foot or in horse-
drawn vehicles. The railways greatly improved London's links
both with its immediate hinterland and with other major cities in
the country, but did not begin to play a major role in London's
internal transport until the underground railways were built and
the built-up area grew to include the city's immediate hinterland.
Within the central city horse-drawn omnibuses remained the principal form of public transport up to the end of the nineteenth
century. The greater part of the railway system was completed
before the end of that century and it is still the system by which
most people get to work in central London. Over 70 per cent of
the 1 075 000 people who arrived daily in central London between
07.00 hours and 10.00 hours during 1975 are estimated to have
travelled by train or underground.(3)

*The Twentieth Century*
While the nineteenth century witnessed great increases in the
mobility of many Londoners, vehicular traffic, whether public or
private, remained the prerogative of the well-to-do. Most people
still lived within walking distance of their jobs, shops and schools
and made use of public transport only on special occasions. In the
early years of the twentieth century the pace of transport developments accelerated. In 1901 London's first electric tramway
opened, from Shepherd's Bush to Acton, and within nine years
the electrified tramway network was carrying almost 800 000 000
passengers a year. The electrifications of the underground was
begun, greatly improving the quality of the air in the system,
and motor buses started to force their horse-drawn competitors
off the road. By 1939 public transport in London was carrying
an estimated 3783 000 000 passengers a year - on average more
than one journey per day per head of the population. Almost 60
per cent of these passengers were carried in London Passenger
Transport Board's buses, while trams, trolleybuses and railways
carried roughly equal shares of the remainder. Services were
greatly disrupted during the war but recovered very quickly, and
by 1948 public transport in London was carrying an all-time peak
of 4675 000 000 passengers a year.

It was a short-lived recovery. Since 1948 public transport has
experienced a continuous decline in the numbers of passengers
carried and a contraction of the services offered. What was once
a highly profitable enterprise must now annually be rescued from
collapse by government subsidies. The reason is not hard to find.
Large numbers of people have deserted public transport for cars.

*The Car*
'The motor vehicle is really demanding a radically new urban
form'. This is the conclusion of the most influential transport
planning document of the 1960s - the Buchanan Report, 'Traffic
in Towns'.(4) It was published in 1963 and exemplifies postwar
transport planning orthodoxy. The rapid growth in car ownership was seen as 'demanding' the provision of more road capacity
to carry it.

In 1959 the Roads Campaign Council held a 'New Ways for London' competition for the best design of a long-term plan for highway development in the city. Buchanan was one of the panel of
judges that awarded first prize to J. A. Proudlove's 'A Traffic
Plan for London'.(5) Proudlove noted that while road traffic was
growing in Britain as a whole at a rate of 14 per cent per year, in
London it was growing at only 2 per cent per year. This relatively
slow growth of traffic in London was seen as the problem to be
solved:

> The present slow increase in traffic will be hastened as
> changes take place to the structure which presently main-
> tains this unnaturally low figure (of 2 per cent) namely the
> lack of road and parking space and the effective public
> transport system able to operate at rates cheaper than
> private travel.

Proudlove's prize-winning scheme for solving this problem involved ambitious changes to London's road structure. Although his
network of urban motorways was not adopted, its guiding principle, that the restraints on London's 'unnaturally' slow traffic
growth should be removed, can be found in the Buchanan Report
and in the major London traffic studies which were subsequently
incorporated in the Greater London Development Plan.

Transport planners frequently illustrate their problems with
'desire line' maps. These lines indicate the numbers of people
estimated to travel between all pairs of zones in the study area.
Figure 4.2, from the London Traffic Survey,(6) shows the pattern
of zone to zone travel in London in 1962 and the growth anticipated, if traffic were not restrained by a lack of road space, by
1981. Figure 4.3 shows the pattern of traffic that was expected
in 1981 if a system of ring and radial roads were to be built sufficient to carry the anticipated traffic.

As the desire lines of car travellers grew longer and wider
during the 1950s, London's main road network, most of which was
established before the invention of the car, came under increasing strain. As more and more Londoners flowed back and forth
across the political and administrative boundaries laid down in a
less mobile era, local traffic authorities found themselves vainly
trying to cope with growing traffic of which an increasing proportion was just passing through their areas. This problem of
traffic management was one important reason behind the creation
of the Greater London Council in 1965.

One of the first official acts of the new council was the announcement of plans to build the 'Motorway Box', the squarish inner ring

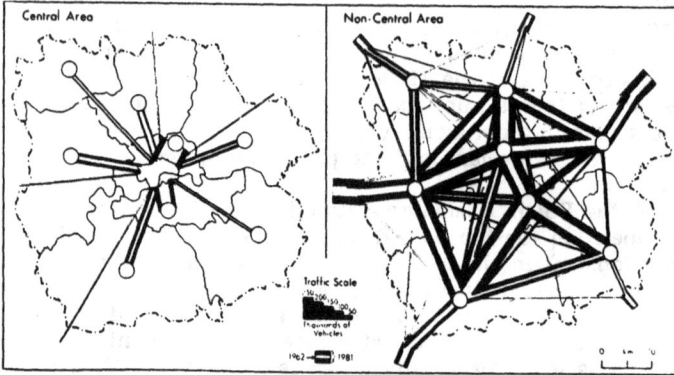

Figure 4.2 Zone to zone travel by road, 1962 and 1981
Source: Greater London Council (1966), 'London Traffic Survey:
vol. 2', p. 129.

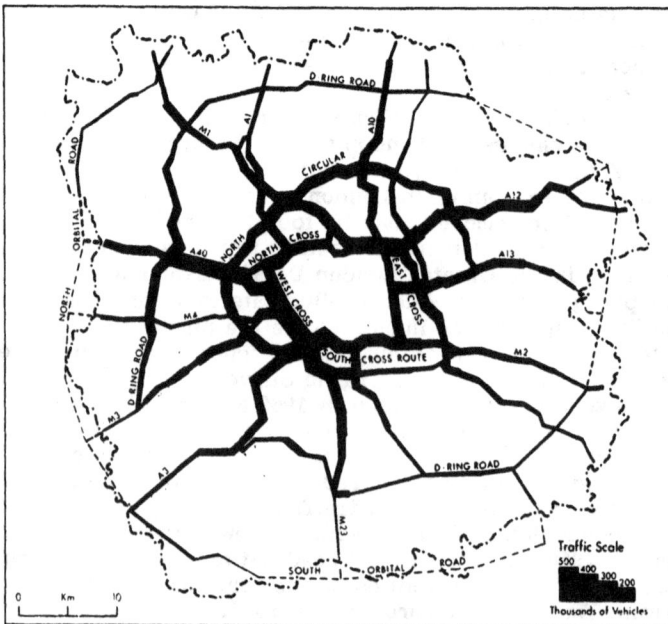

Figure 4.3 'Expected' traffic patterns, 1981
Source: Greater London Council (1966), 'London Traffic Survey:
vol. 2', p. 148

on Figure 4.3, to relieve the congested core of the city. Within a
year it had unveiled the rest of its plans for the vast primary
network of urban motorways that consisted of the additional ring

and radial roads represented by the major flow lines on Figure
4.3.

The plans were extraordinarily ambitious. The Motorway Box,
to be built close to the congested, densely built-up centre of
London, would have been one of the most expensive roads ever
built anywhere. When first presented it was estimated to cost
£25 000 000 per mile. The whole of the proposed network was calculated to cost a total of around £2000 000 000. Inflation quickly
renders such monetary estimates useless as a standard of comparison. The whole scheme would have taken 2500 hectares (6000
acres) of some of the most expensive land in the world and would
have involved knocking down and rebuilding housing occupied by
100 000 people. Its cost would have dwarfed that of other major
transport projects under consideration at the time, such as
Concorde, the Channel Tunnel or the new London airport.

The estimated costs of the scheme increased sharply towards
the centre of the city. The land acquisition costs per kilometre
for the inner Motorway Box were estimated to be fifteen times
higher than for the outer Ringway 3, and construction costs,
because of more crowded working conditions in the inner area,
were estimated to be five times greater.

The plans provoked fierce controversy and became one of the
focal issues of the 1973 GLC elections. As an appreciation of the
disruptive implications of the plans grew, especially in areas
alongside the proposed new motorways, opposition began to organize. This opposition gathered momentum and in 1973 the GLC
Labour Party, having initially proposed the Motorway Box, cam-
paigned against it, and won the election. The new Labour-
controlled council continued to support construction of the outer
ringway, mostly outside the GLC boundary, but within the GLC
area it largely abandoned road building for a policy of restraining
car traffic and promoting public transport. In its 'Transport
Policies and Programme 1977-82' (1976) it declared its intention to
reduce peak hour traffic levels in central London by one-third
below their 1974 levels.

*The Airplane*
The building of the railway network in the nineteenth century and
the motorway network in the twentieth century greatly increased
London's accessibility to the rest of the country. As can be seen
from Figure 4.4, London is the focus of these networks. Since the
war London has also become a major focus of international air
travel. Figure 4.5 shows the growth of traffic through London's
four airports up to a postwar peak of almost 30 000 000 in 1973.
Heathrow to the west of London handled two-thirds of this traffic,
20 300 000, Gatwick to the south took 5 700 000, Luton to the
north-west, 3 200 000, and Stansted to the north-east (which is
used primarily for private flying and test and training flights)
200 000. These four airports together account for over 80 per
cent of Britain's international air traffic.

In 1974 there was a fall in air traffic for the first time since the

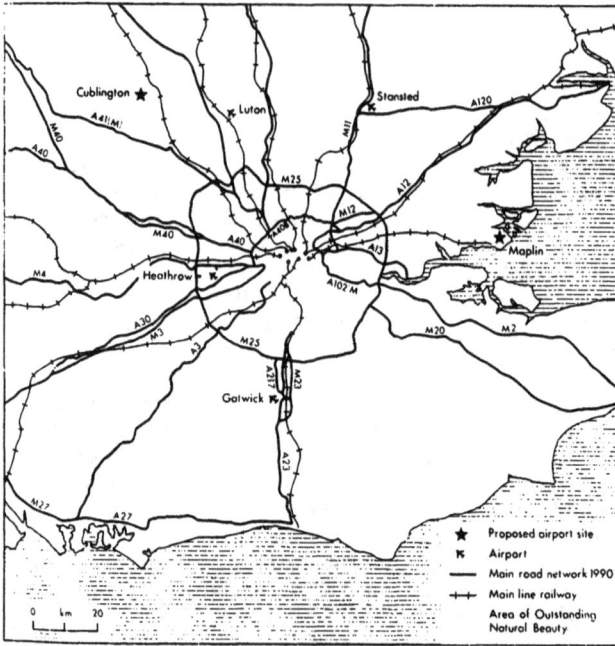

Figure 4.4 Main elements of transportation in the south-east

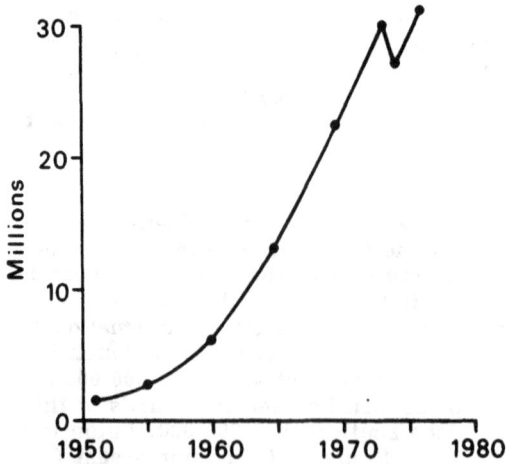

Figure 4.5 Traffic through London's airports
Source: Department of Trade Air Traffic Statistics

war. This has been followed by an uncertain recovery. The principal causes of this abrupt halt were the global economic recession and the very large increase in the cost of aviation fuel in late 1973. Fuel now accounts for nearly one-quarter of total airline operating costs and future traffic trends will inevitably be strongly influenced by its price.

During the period of most rapid growth in the 1960s the air transport planners approached their job with a sense of urgency. Growing traffic was threatening to overwhelm the capacity of the airport system, and resentment was building up in the communities around the existing airports as the noise got steadily worse. The first proposal to deal with this situation was to expand greatly the airport at Stansted. This plan was abandoned in the face of strong local opposition and in 1968 the Roskill Commission was appointed to inquire into the siting of a major new London airport.

The Commission reported over two years later. It predicted that traffic would exceed the capacity of the existing airports by 1980 and recommended that a large new four-runway airport be built at Cublington near Aylesbury in Buckinghamshire.

The Commission attempted to balance the benefits of accessibility against the disadvantages of a major airport. Since London is the focus of south-east England's surface transport system, the most accessible site for a new London airport, from the point of view of its potential customers, would be central London itself. The airport was to perform a central London transport function for which there was no room in central London. The disadvantage of an airport that generally provokes the strongest public reaction is aircraft noise. Sites that would not inflict this on large numbers of people are only to be found at a considerable distance from the centre of the city. The site at Cublington represented the Roskill Commission's idea of a just compromise between these centripetal and repelling forces. But its recommendation met strong opposition from the local inhabitants and the government rejected it and favoured a site on the less accessible and less densely populated Maplin Sands in south-east Essex. But again strong and vociferous opposition was encountered. The government announced a major review of its airport planning strategy and in July 1974 formally abandoned the Maplin project.

Having appeased the opposition at Stansted, Cublington and finally Maplin the government is now left facing the resentful residents of the areas in the noise shadows of Heathrow, Gatwick and Luton who are still clamouring for relief. Although it has increased substantially its estimates of the amount of traffic that the existing airports can handle, it still believes that traffic will exceed this capacity some time before the end of the century. It is conducting yet another inquiry into where new capacity might be built and Stansted, ten years on, is once again a frequently nominated site for a major new airport.(7)

## THE PRESENT DEBATE ABOUT AN UNCERTAIN FUTURE

Since the middle of the nineteenth century London has spread across an ever widening area. While this dispersal has very largely been a consequence of improved transport, it has for most of the period had the blessing of government policy. The various promoters of the railways, the trams and the buses, and most recently the roads' campaigners, have all extolled the benefits of the lower density living conditions that their modes of transport would facilitate. Air travel represents perhaps the ultimate stage of this dispersal, promoting international trade and tourism on an unprecedented scale.

These trends have been actively supported by both central government and the GLC. In recent years the GLC has been a willing participant in the exercise of 'decanting' large numbers of its residents into new towns beyond the Green Belt, while the Location of Offices Bureau advertised the benefits to both employers and workers of migration from London.

The trends have been considered not only desirable but virtually inevitable. Transport planners have consistently based their proposals for new roads and airports on 'unconstrained' traffic forecasts - that is, on estimates of the rate at which traffic would grow if the forces that produced growth in the recent past were to continue into the future unconstrained by lack of road or airport capacity. The current official road traffic forecasts envisage almost twice as much traffic by the end of the century, and the most recent Department of Trade air traffic forecasts foretell a two- to three-fold increase in air traffic by 1990, with potential for considerably more growth by the end of the century.(8)

The belief that such growth will occur is meeting a growing challenge. The sudden large increase in the price of fuel since 1973 which has confounded the trend forecasts is treated in the most recent government forecasts as a unique event that will not be repeated. These forecasts assume that there will be no further increase in the price of fuel this century. The impact of the 'energy crisis' is displayed as a pause on the graphs of future traffic which, after a brief period of readjustment, continue rising as before.

How the price of fuel might behave during the rest of this century is a question that is explored in Chapters 5 and 9. Suffice to say here that if energy becomes much more expensive in the future, many Londoners will become much less mobile than they are at present.

The desirability of achieving the forecast traffic growth is also contested. In the case of air travel, aircraft noise has tended to serve as the focus of the opposition to further growth. Figure 4.6 shows the location of the 55 Noise and Number Index Contour around Heathrow in 1974 when the airport was handling over 20 000 000 passengers. This contour embraces approximately 72 000 people who are considered to have a 'high annoyance rating'. The inner line shows the Department of Trade's estimate of the

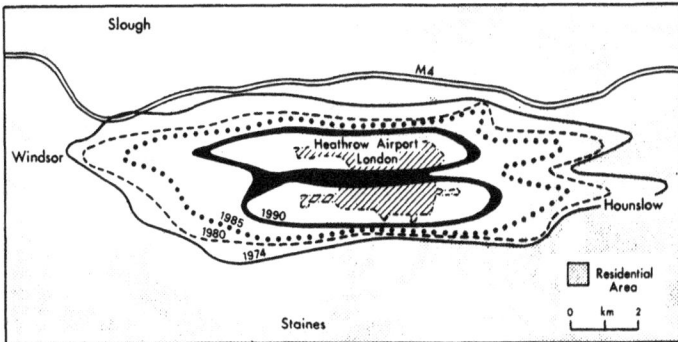

Figure 4.6 Aircraft noise around Heathrow, 1974-90
Source: Department of Trade (1975), 'Airport Strategy for Great
       Britain', HMSO.

location of this line in 1990 when the airport is expected to be
handling 53 000 000 passengers. This reduction in the number of
people estimated to be in the 'highly annoyed' category is impressive but is based upon a contentious assumption: that the average
airplane flying into Heathrow in 1990 will be very much quieter
and very much larger. The assumed growth in aircraft size, up
to a capacity of 800 passengers, is such that the increased traffic
could be carried with no increase in the number of flights. The
interpretation of the 55 NNI contour is also a vexing problem.
Although this contour is intended to represent the outer boundary
of the area of high annoyance, people in the author's household
in Muswell Hill, 19 kilometres beyond this boundary, have not
infrequently been awakened by aircraft noise and have described
their feelings in words that might be roughly translated as 'highly
annoyed'.

However, even if developments in engine technology and air
traffic control were to permit traffic to increase while noise decreased, there are other consequences of the forecast growth that
provoke controversy. Prominent among these is the impact of the
forecast threefold increase in the number of foreign visitors
coming to London. By 1976 the tourist influx was making a pronounced impression on central London; Selfridge's, a prominent
Oxford Street department store, estimated that 40 per cent of its
sales were to foreigners. Presumably three times as many foreigners would spend three times as much money in London's stores
and need three times as many hotels to sleep in, restaurants to
eat in and coaches to travel in. While the tourist industry contem-
plates such developments with apparent satisfaction there is far
from unanimous agreement that the transformation of London that
they would entail should be described as an 'improvement'.

Airports are major sources of direct employment and their attract-
iveness to related service industries and to businesses with foreign
suppliers and customers makes them generators of considerable

additional employment. Heathrow currently employs over 50 000
people and the Department of Trade expects this number to
increase to as much as 65 000 by 1990.(9) For every job at the
airport there is usually another in the immediate vicinity associated with hotels, catering, transport and air freight, as well as
with the provision of goods and services for all these workers
and their families. Beyond this there is the employment that is
associated with internationally oriented businesses for whom
access to an airport is important. In Heathrow's immediate journey-
to-work area, employment directly and indirectly associated with
the airport is approximately one-third of total employment. Since
the residential population of this area is expected to continue to
decrease, substantial growth in airport-related employment must
mean that many workers in the future will have to travel farther
to work. Because the area is already one of labour shortage the
most likely source of additional workers would be to the west of
the airport; it is expected that Heathrow's commuting hinterland
might grow to embrace Reading, 40 kilometres away. The combined
impact of air travellers, commuters and commercial and industrial
traffic places an increasingly heavy burden on the road network
of west London. A recent government report observed that if air
traffic grows as forecast both the roads and London Transport's
underground system serving the airport would become chronically
overloaded without major additions to their capacity.(10)

The present debate about cars is rather more complicated and
confused. While a policy of traffic restraint in the city centre has
been adopted by London, and while similar policies have been
adopted or appear likely to be adopted in most other large cities,
the Department of Transport is still pursuing policies to promote
the growth of traffic on the roads outside these cities. On the
radial roads leading out of the cities these growth and restraint
policies collide head on. In 1958 Buchanan observed, 'the radials
are, in fact, a nasty nagging problem to which there seems to be
no satisfactory answer'.(11) The problem of the radials is now
more nasty and nagging than ever. At the time of writing, this
problem is the focus of a public inquiry into the widening of the
Archway Road, a stretch of the A1, one of the major northern
radial roads leading out of London. The drawing, Figure 4.7,
looking north up the Archway Road from the Archway Bridge,
illustrates the nature of the problem and demonstrates the difficulty of translating the expanded desire lines of the maps in
Figure 4.3 into actual road capacity. Extending the widened road
through the terrace of shops and houses in the photograph for
just a farther 1.5 kilometres, which is what the Department of
Transport is proposing, would cost an estimated £11 000 000 and
destroy the homes of 200 families.

The point where the road narrows is where it ceases to be a
metropolitan road under the jurisdiction of the GLC and becomes
a Department of Transport trunk road. It is ironic that the wide
road belongs to the GLC who enlarged it to an urban motorway
before it adopted its new policy of traffic restraint, while the

Figure 4.7 View northwards along the Archway Road

narrow section is the responsibility of the Department of Transport which is still committed to building roads to accommodate
anticipated large increases in traffic.

Estimates of how much traffic should be accommodated on the
Archway Road have varied enormously. In 1962 the road carried
22 000 vehicles a day. In Figure 4.2, the map showing the predicted unconstrained traffic flows for 1981, it is shown as carrying 180 000 vehicles a day - in 1966 when this estimate was made
only one road in the world, the Harbor Freeway in Los Angeles,
carried more traffic.(12) By 1976 there had been virtually no
increase above the 1962 level because the road and the adjoining
road network were operating at close to their capacities. At the
present inquiry the prediction has been revised by the Department to 75 000 vehicles a day by 1991. But it is not possible to
reconcile even this greatly reduced estimate of traffic growth on
a major London radial road with the GLC's resolve to reduce
traffic in the centre. When confronted by this conflict at the inquiry the Department declared that the GLC's traffic restraint
policies were 'unrealistic' and 'irresponsible'.

The Department and other supporters of the widening scheme
argue much as Buchanan argued in 1958: 'to restrict movement on

the radials would be to squeeze the life out of the town'. They
cite evidence that this is already happening. Since the early 1960s
not only has the dispersal of London's residents continued but it
has been accompanied by a dispersal of jobs as well. Between 1961
and 1974 firms employing 396 000 people either moved out of town
or simply went out of business. By 1975 there were 15 per cent
fewer people entering central London during the morning rush
hours than there were in 1960.

The opponents of the scheme point to the unfortunate example
of many American cities, such as Detroit, whose ambitious urban
road building programmes have manifestly failed to arrest the
decay of their centres. They also note that while the number of
people commuting to work in central London has been declining,
the proportion of commuters who come by car has been rising,
from under 8 per cent in 1960 to over 15 per cent in 1975 (162 000
out of 1 075 000 in 1975). The loss from public transport has been
such that if every car commuter into Central London in 1975 were
to have travelled by public transport, public transport would still
have carried 8 per cent fewer passengers than in 1960.(13) It is
further objected that the scheme is not accompanied by any plans
to relieve the bottlenecks at either end of it, and that such plans,
especially for the constrictions at the inner London end, are
unlikely ever to materialize because they are now recognized to be
prohibitively expensive.

Both sides to the dispute adduce compelling arguments. The
much greater ease of moving about for both cars and lorries is
undoubtedly one of the major attractive forces drawing people and
firms out of London. But providing equally free ease of mobility
within the core of the city could only be achieved by destroying a
large part of it. London's transport problems cannot be treated in
isolation from developments in the wider region of which it is the
focus. The present GLC prescription, of private traffic restraint
and public transport promotion, would seem to stand little chance
of success so long as the Department of Transport is pursuing its
converse policies in the wider region beyond the GLC's boundaries.(14) London is likely to remain a congested island in a sea
of unrestrained mobility. It will increasingly be seen by those who
make the decisions about where to live and work, as an island -
like Venice - dependent on an obsolete transport system. As in
Venice, the tourist trade might prosper, but most other sectors of
the London economy and the people dependent on them, will feel
themselves disadvantaged by their comparative immobility. The
radials are likely to remain a nasty nagging problem until this
policy conflict is resolved.

What the future holds, however, probably depends only in a
small way on the transport policies of the GLC and the Department
of Transport. Looming over the transport policy-makers are forces
beyond their power to influence. The process of change in London's transport that has been described in this chapter has been
influenced only in minor, indirect and often unintended ways by
government policies. If the trend of the past 150 years for trans-

port to become progressively cheaper should be permanently arrested or reversed, then we can expect all the trends that derive from this underlying trend to have their directions altered radically. Today the lament of an increasing number of the least well-off is no longer tha they are 'chained to the spot' but that they are chained to a system of transport that they can no longer afford.

REFERENCES AND NOTES

1  Cited in Barker, T.C. and Robbins, M. (1975), 'A History of London Transport', Allen & Unwin, p. xxvii.
2  White, H. P. (1971), 'London Railway History', David & Charles, p. 77.
3  Greater London Council (1977), 'Transport Policies and Programme: 1978-83', Table 25.
4  Buchanan, C. D. (1963), 'Traffic in Towns', Penguin.
5  Proudlove, J. A. (1960), A Traffic Plan for London, 'Town Planning Review', 31, pp. 53-73.
6  Greater London Council (1966), 'London Traffic Survey: Vol. 2'.
7  Adams, J. G. U. (1971), London's Third Airport, 'Geographical Journal', vol. 137, pp. 468-504, contains a discussion of the debate at the time the Roskill Commission presented its findings.
8  Department of Trade (1978), 'Airports Policy', HMSO, Cmnd 7084.
9  Department of Trade (1975), 'Airport Strategy for Great Britain', HMSO.
10  Ibid.
11  Buchanan, C. D. (1958), 'Mixed Blessing: A Study of the Motor in Britain', Leonard Hill.
12  Greater London Council (1966), op. cit.
13  Greater London Council (1976), 'Transport Policies and Programme: 1977-1982', Table 29.
14  Since this was written the Conservatives have replaced Labour as the majority party on the GLC. They concede the impossibility of accommodating unrestrained traffic growth, but they display considerably less enthusiasm than their Labour predecessors for policies of traffic restraint.

# 5 ENERGY

Mobility requires energy. The mobility transition implies an energy transition. Will the world produce sufficient energy to transport the whole world up to the plateau of high mass consumption and sustain it there indefinitely? There is no agreed answer.

Some answer: yes, of course it will. This is Herman Kahn on the subject:

The real cost of energy supplies has almost always dropped over time; the price of energy, however, fluctuates around this general downtrend according to market conditions. Despite the activities of OPEC and the current pessimism about the extent of petroleum reserves, we believe that energy costs as a whole are very likely to continue the historical downward trend.

There is a general recognition that the oil crisis and subsequent events represented an energy watershed, but it was not a watershed from abundance to scarcity, or even from cheap to expensive, but rather from cheap to inexpensive. That is, many in the industrialized world will continue to drive large cars, if they like them, live in big houses, overheat and over air-condition their homes, expand suburban sprawl, use electric signs and street lighting lavishly, and continue other high-energy consumption activities.(1)

Others answer: no, obviously not:

The best that can be done is to try to discover the upper boundaries of possible production and recognize that actual production can be anywhere below this. Taking such an approach, and looking also at the way humanity is expanding, is dependent upon economic growth and even requires, in many cases, an increasing amount of energy merely to stand still, it becomes obvious that there is not going to be enough energy to maintain present standards of living in the future. The conclusion is unavoidable; too many arguments point in the same direction.(2)

The answer most commonly given by governments everywhere in the world is: yes, we hope so. The following passage from a British Cabinet Office paper entitled 'Future World Trends' is typical of their anxious search for a silver lining in the gathering cloud of energy shortages:

There are some unresolved technological problems, but the scale of world-wide commitment to the thermal reactor, and the pressure for successful development of fast breeder reactors with high fuel efficiency, is so large that there is hope that resources will be found to overcome current or future obstacles.(3)

## FUEL FOR TRANSPORT

Oil is by far the most important fuel for transport. In Britain in 1976 an estimated 99.3 per cent of energy consumed by transport was derived from crude oil.(4) For cars, motor cycles, buses, lorries and airplanes it has, currently, no significant competitors. Also at present, with but few exceptions, these modes of transport have no significant competitors. Of the oil consumed by transport in Britain only the 7 per cent used by trains and ships has any realistic prospect of being replaced by substitute fuels in the short term; conversion to substitutes in the longer term would be very expensive for all modes.

*Table 5.1 Modes of transport: international comparisons*

| Country | Passenger: % of total passenger – miles by each mode in 1975 | | |
|---------|-------------|------|------|
| | *Private car* | *Bus* | *Rail* |
| Great Britain | 81 | 12 | 7 |
| W. Germany | 81 | 12 | 7 |
| Japan | 36 | 16 | 48 |
| USA | 96 | 4 | .4 |

| Country | Freight: % of total ton-miles by each mode in 1975 | | | |
|---------|------|------|----------------|----------|
| | *Road* | *Rail* | *Inland waterway* | *Pipeline* |
| Great Britain | 76 | 19 | 1 | 4 |
| W. Germany | 44 | 27 | 22 | 7 |
| Japan | 71 | 29 | – | – |
| USA | 16 | 40 | 23 | 21 |
| USA(a) | 24 | 36 | 17 | 24 |

Source: 'Transport Statistics Great Britain: 1966-1976', Tables 151-2.

(a) Figures for same year given in Table 1042 of 'United States Statistical Abstract'.

Table 5.1 gives an indication of the difficulties in the path of diverting traffic from modes of transport that depend on oil to modes of transport that have at least the possibility of running on alternative fuels. The figures suggest that countries with dense and concentrated population patterns, such as Japan, offer greater scope for public transport than countries like the United States, but conversely, offer less scope for rail freight. This is

because railways can provide a door-to-door delivery service only
between establishments with their own rail sidings, and the relatively shorter average hauls in Japan, and Britain, are an inducement to use lorries for the main haul as well as collection and
distribution, thereby reducing handling.

In the European countries of the OECD, oil accounted for 64 per
cent of all energy consumed in 1972, and 28 per cent of the oil
was used by transport. The OECD forecasts that by 1985 oil will
account for only 48 per cent of energy consumption but 57 per
cent of this oil will be used for transport.(5) The anticipated
increase in the relative importance of transport as a consumer of
oil is because transport is a 'premium user', that is, it is the user
for which alternatives to oil are the most difficult to find.

Figure 5.1 gives an impression of the magnitude of the task
that the Great Transitionalists have set themselves. It shows the
history of the growth in the production and consumption of oil and
indicates that there is a variety of ways in which present proven
reserves of oil might be spent. If the whole world were to acquire
overnight an American style of energy consumption, proven reserves would last about seven years.

Such an overnight increase will, of course, not happen. Figure
5.1 is but a hypothetical starting point from which to explore the
magnitude of the energy supply implications of a 'successful'
world-wide mobility transition. This transition would take time,
and during this time many things might change.

First, if the mobility transition is to be accompanied by a demographic transition, and it is almost universally agreed that it must
be, the world's population will increase very substantially. For
the Hudson Institute's projections Kahn uses a range of ultimate
world populations from 7500 million to 30 000 million with a most
probable estimate of 15 000; 'Future World Trends' estimates that
the population of the world will grow from its present 4 billion to
8 billion 'at the very minimum'. Thus accommodating population
growth in Figure 5.1 would require increasing the height of the
consumption column by a factor of somewhere between 2 and 8 -
and reducing its width accordingly.

Secondly, the point at which per capita consumption will stop
increasing cannot be assumed to be equal to the present level in
the United States. Consumption in the United States continued to
increase after the 1973 energy crisis and the government remains
committed to economic growth. Kahn estimates that per capita
incomes in the currently developed countries will rise to at least
$40 000 (in constant 1975 dollars) before they stop growing, com-
pared to a per capita income in 1976 in the United States of less
than $7000.(6) If oil consumption were to increase proportionately
then the consumption column would have to have its height increased a further six- or sevenfold and, again, its width reduced
by the same factor.

But thirdly, countering the first two changes, are anticipated
improvements in the efficiency with which energy is used. 'A Low
Energy Strategy for the United Kingdom' (1979) contains a survey

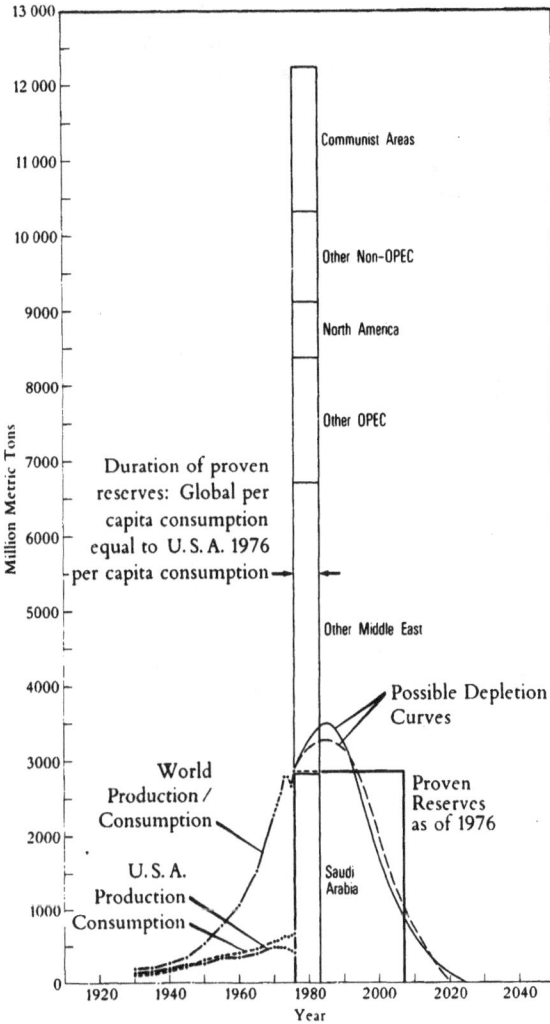

Figure 5.1 Oil Consumption
Note: The rectangle labelled 'proven reserves as of 1976' indicates
how long proven reserves would last at the 1976 rate of consumption. The tall thin column illustrates how long these reserves
would last if the whole world were to consume oil at the rate at
which it was consumed in the United States in 1976. This column
is divided into portions indicating the regional distribution of
proven reserves.
Sources: 'UN Statistical Yearbooks'; 'Statistical History of the
        United States from Colonial Times to 1970'; 'Scientific
        American', vol. 238, pp. 42-9.

of the most likely of these improvements and the savings that
might result if they were widely implemented. The list for transport is quite impressive: streamlining of cars, reduced tyre drag,
vehicle weight reduction, improved lubrication, more efficient
driving techniques, better maintenance, electronic controls,
thermostatically controlled fans, variable transmissions, more
efficient engines, regenerative braking for trains, fluidized-bed
coal burning turbogenerators to replace diesel engines, anti-drag
devices for lorries, higher load factors for lorries, fluidized-bed
coal burning engines and improved hull designs for ships, high-
bypass turbofan engines and low-drag supercritical wings for
aircraft, and the development of high-energy density sodium-
sulphur batteries. On the basis of detailed assumptions about the
progress that might be made on all of these fronts, 'A Low Energy
Strategy' concludes that it would be possible for car, van and
motor cycle traffic to increase by 80 per cent, air traffic by 200
per cent, and road freight by 95 per cent by the year 2035 while
keeping energy consumption at 1976 levels.(7) More generally, it
is argued in 'A Low Energy Strategy' that the United Kingdom
could treble its gross domestic product by 2025 without increasing
its consumption of energy.

Herman Kahn, widely regarded as a technological optimist,
assumes for his projections of energy consumption a fourfold
increase in the next two centuries in the efficiency with which
energy is produced, converted and utilized. For his central projection, which represents the net effect of the operation of all the
above factors, he assumes a 3.75-fold increase in world population,
a fifteenfold increase in world per capita incomes, and a 14.4-fold
increase in world energy consumption.(8) Thus, if oil were to
maintain its present share of the world energy market, while this
market expanded more than fourteenfold, the column representing
annual consumption in a world of universal affluence would have
to be more than three times higher, and thinner, than in Figure
5.1. In other words, the great transition implies consumption of
oil, or substitutes for oil, at a rate, sustained indefinitely, that
would exhaust the world's proven reserves in about two years.

The world's proven reserves do not of course represent all the
oil there is, only that which oil companies are fairly confident of
extracting profitably from specific places with currently known
extraction technology. There may well be much more. The report
of the Workshop on Alternative Energy Strategies, a study involving the governments of fifteen major non-communist oil-importing
countries, suggests that ultimately recoverable, but as yet un-
proven, reserves could be three to five times greater than currently known reserves.(9) In this case the ultimate sustained rate
of consumption could take as many as ten years to exhaust global
oil reserves.

There may well be *very* much more. Since the time of Malthus
many forecasts of impending failures of supply to match growing
demand have been disproved. They have been disproved because
they have been based on technologies known at the time, or on

anticipations of technological advance that turned out to be excessively cautious. They have been primarily failures in technology
forecasting rather than failures in estimating quantities of physical substances. Furthermore, there is reason to question the
motives of some energy forecasters. Those who are best qualified
to predict how much oil might be found in the world are people
with long experience of the oil business. The oil business is
chronically dissatisfied with the proportion of its profits that
governments take in tax. An impending shortage of oil constitutes
a powerful argument for tax relief to permit more investment in
exploration and development. Because oil forecasters either work
for, or are crucially dependent on data supplied by, major oil
companies there is reason to suppose that most forecasts of reserves will err on the side of caution. Finally, energy forecasts
are not immune to political pressures. The American Central
Intelligence Agency has produced its own estimates of oil reserves.
With its unrivalled resources for gathering and interpreting information it should be in an unrivalled position to estimate these
reserves realistically. But its published, and much publicized,
prediction of impending scarcity has come under suspicion of
being an attempt to fabricate an 'energy crisis' for purposes of
winning public and congressional support for President Carter's
energy policies.(10,11)

Nevertheless, Figure 5.1 illustrates the ephemeral nature of oil
as a source of energy in the context of projected global energy
consumption aspirations. It is possible by altering the economic,
demographic, technological and geological assumptions discussed
above, to extend by a few years the life of oil as a major source
of energy. But oil clearly has a severely limited future; as it
becomes scarcer it will increasingly be reserved for premium uses
such as transport, but ultimately, by the early part of the next
century at the latest, substitutes will have to be found. The
debate, at the moment, is focused on the question of whether they
are likely to be found in time to permit the uninterrupted progress
of the Great Transition. The crucial importance of energy supplies for this progress can be gauged by the impact made on the
world economy by the 1973 pause in the rising graph of Figure 5.1.
The rising graphs of the economic and mobility transitions must
be accompanied by the energy graph. They cannot continue to
rise indefinitely if the energy graph falls away, or even merely
levels off. Herman Kahn's Hudson Institute is adamant in its
optimism that it will continue to rise: 'The basic message is this:
Except for temporary fluctuations caused by bad luck or poor
management, the world need not worry about energy shortages or
costs in the future.'(12)

For the immediate term, it argues, reserves of coal provide
adequate insurance: 'If oil in the Persian Gulf were to disappear
over the next 10 years, the relatively immediate and obvious need
for coal would dictate that the adjustment be made swiftly and
effectively.'(13)

And in the longer term, according to this 'scenario', science and

technology will join forces to provide the world with abundant
and inexhaustible supplies of energy. Table 5.2 contains two
summaries of the major possible future energy sources. But there
is reason to doubt that these resources can be tapped swiftly and
effectively. The failure of America's Project Independence to prevent an increase in American oil consumption after 1973, or to
prevent an increased dependence on foreign suppliers for this oil,
from 36 per cent in 1973 to almost 50 per cent in 1979, calls into
question the possibility of a smooth transition to alternatives to
oil. (14)

*Table 5.2  Probable reserves as a multiple of probable oil reserves*

| Source | Hudson Institute[a] | World Energy Conference[b] |
|---|---|---|
| Natural gas | 1 | .5 |
| Tar sands and | | |
| Oil shale | 2–133 | 2.7 |
| Coal and lignite | 13 | 3.9 |
| U–235 | 200 | 8.2 |
| Uranium for breeder | | |
| reactors | 〉6666 | 16.5 |
| Fusion reactors | 〉66 million | — |
| Solar radiation | 2 per year | — |
| Ocean gradients | 1.3 per year | — |
| Organic conversion | .08 per year | — |
| Geothermal | 〉66 million | — |

(a) Kahn, H. et al. (1976), 'The Next 200 Years', Table 6, p. 82.

(b) Presented by White, N.A. (1977), in The Cost of Energy over the Next Decade, in 'Glass
Technology', Table 3, p. 59.

But still, it would be a foolish person who would deny the possibility of scientific and technological breakthroughs such as
those envisaged in Table 5.2; the fact that fusion reactors do not
exist does not preclude the possibility of ways being devised to
produce, contain and utilize the temperatures in excess of 100
million °C involved in the fusion process. Very few people are
qualified to hold an independent opinion on such questions.
Equally the evidence of the way in which industrial societies can
channel their productive capacities in times of war means that one
cannot rule out the possibility of rapid adjustments on a heroic
scale once sufficient numbers of people become convinced of the
urgency of a problem and become committed to a solution to it.


## THE MORAL EQUIVALENT OF WAR

On 18 April 1977, in a nationally televised address, President Carter attempted to impart to the nation a sense of urgency about the

energy problem. He warned that 'national catastrophe threatened the United States if it did not alter its energy habits radically; sacrifices and changes would be required 'in every life'. The changes required he outlined in his energy programme. It would, he said, involve every citizen in 'the moral equivalent of war'. The central objective of this war, outlined in the former President's programme, was the reduction of the growth rate of the country's energy consumption to 2 per cent per year.

If such a 'moral victory' were to be achieved and sustained it would be unlikely to bring about a durable peace. Keeping the growth rate down to 2 per cent a year in the country that consumed more than a quarter of the world's oil in 1976 would over the next 200 years result in a more than fiftyfold increase in that consumption. The United States's government obviously does not plan so far into the future. But the fact that such a substantial growth rate can be considered a moral victory in the energy war is an indication of the momentum of energy consumption patterns in the industrialized world, and of the mounting pressure to develop alternatives to oil. If world energy consumption is to continue to rise to sustain a growing world economy, the war to bring energy supply and demand into balance is not one in which final victory can ever be achieved. It is a war against diminishing marginal returns; its intensity will increase for ever more.

The suppliers of the world's energy wield power without historical precedent. The world's largest company, Exxon, had annual sales in 1976 of more than $42 000 million, a figure greater than the gross national product of all but the wealthiest countries in the world. The ten largest companies, with the exception of Ford and General Motors, are all oil companies; together they had sales of over $240 000 million, more than the gross national products of all but five countries. Of the fifty largest companies in the world, the major activity of eighteen is the production of oil, and the major activity of a further seven is the production of cars and trucks, the principal consumers of oil. (15) These multinational companies have demonstrated that they have sufficient power to defy both national governments and the United Nations. In the much publicized case of Rhodesia, to cite a recent example, the exercise of this power in defiance of an international oil embargo sustained the white minority government in power for over ten years.

Not only is the production and sale of the life blood of the world economy dominated by a small number of companies, it is dominated by an even smaller number of countries. The central column of Figure 5.1 illustrates the concentration of known reserves in a few sparsely populated countries and sheikdoms in the Middle East. Since 1973 there has been a significant shift in the control of these supplies, and of the profits derived from them, from the major oil companies who developed them to the rulers of the countries that possess them. The result has been the concentration of incredible wealth in very few hands, and a situation that is economically, politically and militarily extremely unstable.

The vital dependence of the economy of the non-communist world on these precarious supplies has made the Middle East the focus of world diplomacy. The cutting off of these supplies would almost certainly provoke not the moral equivalent of war, but war itself. The following passage, taken from an article in the American current affairs quarterly 'The Public Interest', illustrates the danger:

> The only serious security threat is the prospect of Soviet intervention in the Gulf, whether by direct military incursion or via client regimes. The fact is that the remaining oil and gas resources of the Persian Gulf have a present market value of some $16 trillion. The U.S. can no more afford to have these fall into Soviet hands than it can afford to lose Western Europe. The rate of current Persian Gulf production or the level of U.S. supply dependence makes absolutely no difference to the strategic importance of that region of the world. Were the U.S. somehow to achieve 100 per cent energy independence, it would still have to counter overt Soviet aggression in the Persian Gulf with strategic power.... The task of meeting that threat has belonged to the defence establishment all along.(16)

The curious denial of any connection between the strategic importance of Gulf reserves, and the dependence on them of the world's largest consumer of oil, stems from a failure to appreciate that their market value is a reflection of both their economic and their military significance. So long as the world continues to grow more dependent upon sources of energy that have a very uneven geographical distribution, these sources of energy will become the object of increasingly intense competition, and will be more likely to provoke military conflict.

All anticipated alternatives to crude oil that might be developed on a scale sufficient to sustain global economic growth, and a concomitant growth in mobility, would involve similar economic and geographical concentrations of control and would have similar military implications.(17) Although there are possibilities for making more effective use of diffuse energy sources such as solar radiation for low grade uses such as domestic heating, all currently known alternatives to crude oil as a fuel for transport are more expensive than crude oil or require more advanced technology, or both. The alternatives are essentially two: oil derived from coal, shale or tar sands, and fuels, either electricity or hydrogen, produced by nuclear fission or fusion.(18)

The first set of alternatives would have, on a growing scale, all the same environmental costs associated with their distribution and consumption that oil now has. The costs of production, both economic and environmental, would be very much higher.

Resources of oil from shale and tar sands, usually figure as reassuringly large numbers in inventories of future world energy sources. The following passage from the OECD's 'Energy Prospects to 1985' gives an indication of the scale of the processing activity that would be involved if oil shales, the most prom-

ising of the two sources, were ever to replace crude oils as a significant energy source:

the production of 1 million barrels per day [one sixteenth of
the United States rate of consumption in 1976] of oil from
shale by *surface* processing would require the mining of
approximately 570 million tons of oil shale annually. This
approximates the annual level of United States coal production and therefore raises severe doubts as to the
availability of mining personnel and equipment. Additionally,
surface processing of oil shale requires great volumes of
water (about one-third of which is used for cooling purposes
and two-thirds to solidify the spent oil shale): it has been
estimated that development of an oil shale industry to a level
of 3-5 million barrels per day (surface processing) would
require essentially all of the available water in the region
concerned.... Finally and perhaps most important, the disposal of spent oil shale, the volume of which grows during
the mining process, presents a tremendous environmental
problem.(19)

Although in situ conversion of the oil shales to oil or gas is a
theoretical possibility, and might obviate many of these costs, it
would require the generation perhaps by nuclear explosion of heat
underground on a vast scale. It is, as yet, not a practical alternative and would have environmental costs that can only be
guessed at if conducted on a scale sufficient to make it a major
substitute for oil. Known reserves are even more highly concentrated geographically than those of crude oil, Canada and the
United States have almost 90 per cent,(20) and the associated
problems of highly concentrated ownership of a global resource
would be at least as great as those now experienced with oil. It
would appear most unlikely that either country would wish to bear
the burden of environmental costs necessary for export production
on a scale that would make a significant impression on the projected global demand for oil.

Similarly with coal. Although it is possible to produce oil from
coal, the only serious attempt to do so, in South Africa, indicates
that it requires about 1 tonne of coal to produce .9 barrels of oil.
(21) Thus, satisfying the United States's 1976 oil requirements
from coal would require 6772 million tonnes of coal, or more than
ten times the quantity of coal mined in that year. Although coal
reserves are more evenly distributed than those of oil, two countries, the United States and Russia, are variously estimated to
have between 58 and 84 per cent of the world's reserves. The
devastation caused by the mining, let alone the process of converting to oil, of sufficient coal to satisfy these countries's own
requirements again makes it unlikely that they would be interested
in producing for export.

Raw materials for the nuclear alternatives are much more evenly
distributed about the world than reserves of fossil fuels. But the
economic and technological resources required to exploit them are,

at present, dominated by a small number of countries. Such is
the lethal potency of the processes and materials involved in
nuclear energy that this dominance is jealously guarded.

In 1976 transport in Britain consumed, almost entirely in the
form of oil, more than fifteen times as much energy as was generated by nuclear power.(22) Thus for nuclear power to become a
substitute for a significant part of the market now served by oil
would require a very large increase in nuclear generating capacity. It would also require a very large increase in the resources
devoted to the security of the nuclear industry.

Effective security depends on surveillance and secrecy, on
learning as much as possible about the enemy while permitting the
enemy to learn as little as possible about oneself. The requirements of security are in direct and inescapable conflict with the
requirements of a free and democratic society. As the stock of
toxic nuclear substances increases, the problem of keeping them
out of the 'wrong' hands is increasingly adduced as an argument
for concealing information about the means of security. As dependence on nuclear power grows, the less admissible becomes
discussion at public inquiries of the wisdom of the government
policies that foster this dependence. (See Chapter 12 on the problem of discussing policy at inquiries.) As the threat of nuclear
proliferation grows, the larger the area of discussion subsumed
by the prohibition 'national security'. As cost over-runs grow,
the more reluctant become those responsible for the miscalculations
to provide the information necessary for an informed discussion of
the reasons. As the quantity of low-level radioactive emissions
increases, the stronger becomes the argument that public discussion of their uncertain consequences will spread panic among
the scientifically untutored masses. The greater becomes the
dependence of the world on nuclear power, the smaller becomes
the possibility of the lay public participating in informed discussions about it.(23)

The growth of the nuclear power industry has been accompanied by a growth in the opposition to it. Those who do not wish
the industry well range from pacifists to terrorists and come in a
great variety of political colours. Those who represent the most
serious threat to security are the least likely to parade in their
true colours. Hence effective security requires keeping an eye on
all opposition. Spokesmen for the industry are most anxious to
assure law-abiding citizens that the security requirements of the
industry pose no threat to their civil rights. P. J. Searby, Secretary of the United Kingdom Atomic Energy Authority, has
offered the following reassurance in a letter to 'The Times':
'Bodies and individuals opposed to the development of nuclear
power would not be subject to security surveillance unless there
was reason to believe their activities were subversive, violent or
illegal' (12 August 1977).

And F. J. Chapple and J. Lyons, respectively Chairman and
Secretary of the Electricity Supply Industry Employees' National
Committee, in another letter to 'The Times' dismiss concern about

civil liberties as scare mongering:

> This is an aspect of the nuclear debate which needs airing
> but Mr Seighart [Joint Chairman of Justice] fails, it seems
> to us, to explain why the existence of a number of fast
> breeder reactors would create any more of a problem in
> respect of civil liberties than does the present existence of
> the Ministry of Defence and all its range of activities (27
> September 1977).

Those who are attempting to subvert the prevailing belief that
the growth of the economy and energy consumption can and should
continue indefinitely are little comforted by the assurance that
only the subversive are to be targets of surveillance. Nor are they
reassured by the suggestion that nuclear security poses no more
of a threat to civil liberties than does the Ministry of Defence.
Civil liberties end where the realm of the military begins. Armies
are run as dictatorships. The cardinal virtue of the soldier is
obedience. He has no right to withdraw his labour if he disagrees
with his general about the identity of the enemy. The operations
in which he engages, even in peacetime, are shrouded in secrecy -
from the enemy, from subversives, from the civilian population
and, usually, from himself. His right to privacy is sacrificed to
the demands of security. Passes, security checks and surveillance
are normal features of military life. Attempts to subvert the dogma
of the high command by reasoned argument are ruthlessly sup-
pressed.

A very plausible account of the threat to civil liberties posed by
nuclear power is contained in a report entitled 'Nuclear Prospects'
(24) The report describes some of the catastrophic consequences
that could follow from a failure of nuclear security in the plutonium fuel cycle. They range from releases of radioactivity from
nuclear electricity generating plants, to the havoc that could be
wreaked by psychopaths armed with plutonium or even atomic
weapons. It then describes what is known of present security
practice. Its salient features include positive vetting for all the
industry's professional staff (i.e. intensive scrutiny of their
characters, associations and vulnerability to 'subversion'), a
special armed constabulary with powers to engage in 'hot pursuit'
and to arrest on suspicion, and an apparatus of surveillance - the
nature an extent of which is an official secret.

Placing credible threats of sabotage or malicious use of stolen
nuclear material in the context of present security practice, the
report proceeds to outline the probable security requirements of
the greatly expanded nuclear programme envisaged by the industry's proponents. The picture that emerges is one of a society
crucially dependent for its very existence on electricty generated
in heavily defended installations to which only vetted people are
permitted entry, supplied with fuel transported under armed
escort, from equally well guarded fuel refining and reprocessing
facilities. Given the scale of the operation, the numbers of people
involved, the widespread hostility it would be bound to provoke,
and the extreme consequences of a failure of security, the report

argued that it would be irresponsible for the nation not to guard it with a strict, pervasive, militaristic security service.

The authors of the report, Michael Flood and Robin Grove-White, anticipated that those responsible for nuclear security might resent their 'casting a light on activities which, arguably, rely on darkness for their full effectiveness'. They cited an article in 'Atom' by Sir John Hill (Chairman of the UK Atomic Energy Authority) in which he condemned a similar exercise conducted in the United States on the grounds that it 'provided a great deal of information which might just give the necessary encouragement to terrorists contemplating some nuclear outrage'. But they remained unrepentant: 'It would be far the unhappiest and most distinctive feature of nuclear power if its successful development were held to involve hazards so great that a democracy could be prohibited from talking about them.'

When they proffered their report to the electricity supply industry for comment they had their anticipation confirmed. They were told by one of the most senior figures in the industry that their report was 'seditious'. Nowhere did their report advocate, or even hint at condoning, the violent overthrow of the government. On the contrary their case against the nuclear industry rested on the argument that its further expansion would do violence to cherished democratic institutions. They argued that the expansion of an industry inherently incapable of democratic control would foster an opposition that had a diminishing regard for the conventions of democratic protest.

The inability, or unwillingness, of the industry's leadership to distinguish between sober, precautionary argument, and incitement to the violent overthrow of the state, illustrates the inherent totalitarian tendencies of the nuclear power industry. An infallible sign of a tyrant is his inability to distinguish legitimate opposition from illegitimate - all opposition is sedition. The principal difference between the security services required by a society dependent on nuclear power, and those by a country at war against a foreign enemy, is that the former must be directed against threats that are largely internal. Waging the moral equivalent of nuclear war on the energy crisis would involve living with the moral equivalent of a security system appropriate to a society in the throes of civil war.

The belief, held by some, that the world can and will yield sufficient energy to sustain a global mobility transition rests upon discoveries, scientific and geological, that have yet to be made. Such discoveries, if they were made, could only be exploited at the greatly increased risk of war, or its moral equivalent.

## REFERENCES AND NOTES

1   Kahn, H. et. al. (1976), 'The Next 200 Years', William Morrow & Co., pp. 58-9.
2   Foley, G. (1976), 'The Energy Question', Penguin, p. 327.

3   Cabinet Office (1976), 'Future World Trends', HMSO, p. 13.
4   Leach, G. et. al. (1979), 'A Low Energy Strategy for the United Kingdom', International Institute for the Environment and Development, Science Reviews, p. 137.
5   Organization for Economic Co-operation and Development (1977), 'Energy Problems and Urban and Suburban Transport', Table 1, p. 11.
6   Kahn (1976), op. cit., pp. 54-7.
7   Leach (1979), op. cit., pp. 155-69.
8   Kahn (1976), op. cit., Table 3, p. 62.
9   Flower, A. R. (1978), World Oil Production, 'Scientific American', vol. 238, no. 3.
10  'The Guardian', 11 April 1979, p. 10.
11  Stockman, D. A. (1978), The Wrong War? The Case against a National Energy Policy, 'The Public Interest', no. 3, p. 32.
12  Kahn (1976), op. cit., p. 83.
13  Ibid., p. 60.
14  An indication of the lack of seriousness with which Britain considers the energy crisis is provided by the fact that in 1979, in the midst of all the publicity surrounding petrol shortages and the Iranian revolution, May car sales reached an all-time peak.
15  Lindsay, J. (1976), 'The International Enterprise - A Framework for Discussion', The Administrative Staff College, Henley, Appendix 1.
16  Stockman (1978), op. cit., p. 44.
17  One alternative energy strategy that does not have such implications is the solar strategy advocated by Barry Commoner. The principal fuel in this strategy would be methane generated from biomass, supplemented by heat from solar panels and electricity from windmills and photovoltaic cells. Clearly these sources must become more important as supplies of non-renewable fuel are depleted. But they require enormous areas of land for the collection of solar energy, and none of their proponents suggest that they are likely to be able to supply energy on anything approaching the scale implicit in the global transitions discussed in earlier chapters. A good summary of the limitations of solar energy can be found in Foley (1976), op. cit., Chapters 11 and 12.
18  Kahn (1976), op. cit., p. 65.
19  OECD (1974), 'Energy Prospects to 1985', p. 99.
20  White, N. A. (1977), The Cost of Energy over the Next Decade, 'Glass Technology', 3 June, p. 59.
21  Foley (1976), op. cit., p. 241.
22  Calculated from Leach (1979), op. cit., Tables E2 and E11. This is a very rough and ready comparison. It ignores the fact that electricity can be converted into motive power more efficiently than oil. A more precise indication of the magnitude of the task of meeting transport energy requirements with electricity would require a large number of assumptions about feasible generation, transmission and storage efficiencies.

23  A good account of the largest public inquiry ever held in
    Britain on the question of nuclear power is contained in
    Breach, I. (1978), 'Windscale Fallout', Penguin.
24  Grove-White, R. and Flood. M. (1976), 'Nuclear Prospects',
    Friends of the Earth, 9 Poland St, London W1.

# 6  EQUITY

As long as motorcars were few in number, he who had one
was a king: he could go where he pleased and stop where
he pleased; and this machine itself appeared as a compen-
satory device for enlarging an ego which had been shrunken
by our very success in mechanization. That sense of freedom and power remains a fact today only in low-density areas,
in the open country; the popularity of this method of escape
has ruined the promise it once held forth. Lewis Mumford (1)

Estimates of the number of cars in China in the mid-1970s range
from 28,900 to 85,000. In 1972 according to the FBI's *Uniform
Crime Statistics* there were 33,283 cars stolen in Detroit,
37,615 in Boston, 43,083 in Chicago, 67,794 in Los Angeles,
and 80,346 in New York City.(2)

Implicit in the judgment that all the graphs of increasing mobility
presented in Chapter 1 represent progress, is the idea that the
mobility transition is a democratic process - ultimately all people
everywhere will climb the ladder of progress and share equally in
the benefits of high mobility. According to this view, present
disparities in levels of mobility are justified by the argument that
progress cannot occur everywhere simultaneously; it must start
somewhere, and it takes time. The wealthy and powerful are the
first to gain access to new transport technologies, but this, it
is argued, does not make the new modes of transport elitist, at
least not permanently. The mobility transition is a process in
which a few forge ahead in the initial stages to be followed by a
universal levelling-up in the final stages. Evidence already examined in earlier chapters suggests that internationally this levelling-up is unlikely ever to happen. This chapter examines the way
in which the distribution of the benefits of mobility has been
changing within countries that are close to the top of the mobility
ladder.

## TIME-SPACE DOMES (3)

The principal and most obvious benefit of increased mobility is
the extension of the range of opportunities both economic and
social that it provides its beneficiaries. The price paid for this
benefit is less obvious.

Figure 6.1 illustrates the way in which the opportunities of which
people avail themselves by travelling are typically distributed over

space. It illustrates the fact that the frequency with which people make journeys has an extremely strong tendency to decrease as journey length increases. When the journeys made by a number of people are plotted on a graph as though they had a common origin, and with the position of each dot indicating the length and compass direction of each individual journey, a clear distance decay effect emerges. The density of the dots decreases with distance from the centre, permitting the generalization of the pattern as a domed 'mobility surface'.

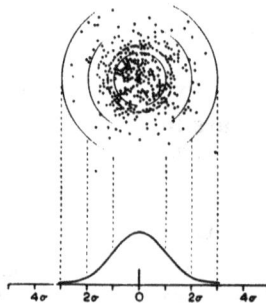

Figure 6.1 A centred interaction field. Broken ring shows mean radius of interaction.
Source: Adams, J. G. U. (1971), London's Third Airport, 'Geographical Journal', vol. 137, pt 4.

The dots in Figure 6.1 represent the travel behaviour of a group of rats. Inferences drawn about human behaviour from the behaviour of rats can be misleading, but in this case, however, nothing need be assumed about the motives of the travellers involved. The figure illustrates certain inescapable principles that apply to the behaviour of all species of traveller, including Homo sapiens.

For purposes of appreciating some of the consequences of current mobility trends we can think of the mobility surface as a 'time-space dome' within which people spend their lives. The height of the mobility surface at any particular point is proportional to the amount of time that is spent at that point. The volume of the dome corresponds to the total amount of time that people have to spend. They can alter the shape of the dome, but not its volume. People and societies that do not travel much inhabit high, rigid, confined domes, those who travel a great deal live in low, flexible, spread-out domes. But they all live within domes of the same capacity because they all have the same number of hours a day at their disposal.

Table 6.1 describes recent trends in mobility in the United States and Great Britain. It will be seen that in the twenty-five years between 1960 and 1975 the number of miles travelled per person per day in the United States almost doubled, to 27.4 (43 kilometres), and that in the twenty-three years between 1954 and 1977 this number more than doubled in Britain, to 15.3 (24 kilometres). Thus although people in the United States, on average, are still almost twice as mobile as people in Britain, the gap is closing.

*Table 6.1  Passenger travel*

| United States[1] | | Miles travelled per person per year | | 1975 as a percentage of 1950 |
|---|---|---|---|---|
| | | *1950* | *1975* | |
| Car[2] | | 4567 | 8987 | 196% |
| Surface public transport | | | | |
| Intercity bus | 172 | | 132[3] | 77% |
| Intercity train | 212 | | 47[3] | 22% |
| Urban transit[4] | 570 | | 164 | 29% |
| | | 954 | 343[3] | 36% |
| Airplane | | 66 | 686[3] | 1039% |
| | | 5587 | 10016 | 180% |
| Miles per person per day | | 15.3 | 27.4 | 180% |

| Britain[5] | | *1954* | *1977* | 1977 as a percentage of 1954 |
|---|---|---|---|---|
| Car, taxi, motor cycle | | 961 | 4234 | 440% |
| Surface public transport | | | | |
| Bus | 1024 | | 606 | 59% |
| Train | 493 | | 389 | 79% |
| | | 1517 | 995 | 66% |
| Airplane[6] | | 32 | 365 | 1141% |
| | | 2510 | 5594 | 222% |
| Miles per person per day | | 6.9 | 15.3 | 222% |

Sources: 1 'Statistical Abstract of the United States 1977', US Bureau of The Census, Tables 1043 and 1056; 2 Based on Table 1056. Assumes 2.4 persons per vehicle in rural areas and 1.4 in urban areas; 3 Figures for 1974; 4 Based on numbers of passenger journeys, journey lengths assumed to be the same as the average journey on London Transport in 1974. i.e. 5 miles; 5 'Transport Statistics Great Britain 1964-1974' and '1967-1977', HMSO, Tables 9 and 11 respectively; 6 Based upon scheduled air transport services flown by British operators, 1954 figure from 'Annual Abstract of Statistics', 1956;

Figure 6.2 gives an indication of the dramatic transformation that takes place in the time-space dome of a society as its level of mobility doubles and redoubles. The high peaked curve represents the space-time dome of an individual, or a group, with an average trip length of 1.25 miles (2 kilometres). The other curves represent

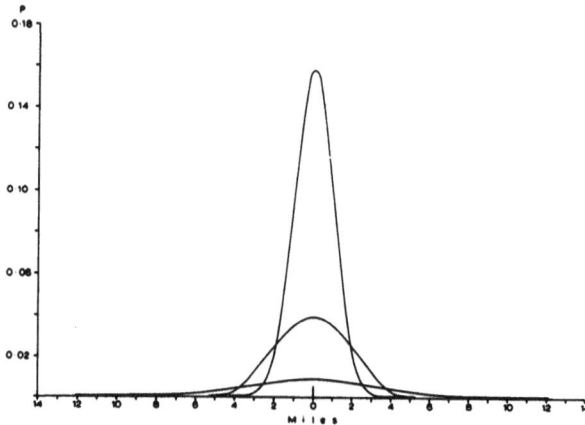

Figure 6.2 Variations in intensity of interaction with changes in mean radius.
Source: Adams, J. G. U. (1971), London's Third Airport, 'Geographical Journal', vol. 137, p. 4.

the effect of doubling and then doubling again the average trip length. The new, more remote, opportunities of which people avail themselves when they become more mobile are generally not additional to those previously enjoyed, but substitutes for opportunities previously taken closer to home, and now foregone. Figure 6.2 illustrates the obvious fact that if people, in spending the time at their disposal, distribute themselves more widely over space, the amount of time they spend close to home, and perhaps at home, must be substantially reduced. Although the travel behaviour of people will rarely display a pattern as simple and symmetrical as that of rats, the essential principle illustrated by the time-space dome is inescapable: if people in their travelling choose to spread themselves more widely, they must spread themselves more thinly.

CHANGE AT THE CENTRE

The height at the centre of the surfaces in Figures 6.1 and 6.2 is proportional to the amount of time spent at home. Evidence relating to changes in the amount of time that people are spending at home is rather sparse and contradictory.(4, 5) But Figures 1.6 and 1.7 of Chapter 1, which chart the increase in the ownership of telephones and television sets in countries such as the United States and Britain, suggest that whatever the trend in the amount of time spent at home, social life there is being sapped electronically.

    Although interaction by telephone is a weaker form of communication than that afforded by face-to-face contact, the pattern of growth is similar to that presented by the physical interaction indices. Growth is extremely rapid and fastest of all at the greatest distances. In 1955 in Britain 7.8 per cent of all calls were domestic long-distance calls and 0.04 per cent were international; by the year 2000 it is forecast that 22.17 per cent of all calls will be long distance and .75 per cent will be international.(6) The impact of television is the subject of endless debate. It is estimated that in the United States, the country with the highest ownership rates, the average person spends over four hours a day watching television.(7) It cannot be seriously doubted that one of the consequences of having a television set on for more than four hours a day is to attract the attention of those in the household away from each other and towards the material presented. Television coverage of current events now extends, intermittently, depending on the 'newsworthiness' of events, to almost all parts of the world. If graphs were to be drawn to describe changes in an average individual's centred information field, the spreading and flattening effects caused by telecommunications would be even more pronounced than the changes depicted by the graphs of physical interaction. If the growing amount of time spent interacting electronically is charged against time spent at the centre (i.e. those so interacting are counted present in the flesh but absent in spirit) then Figure 6.2 can be considered a valid representation of the nature of change at the centre as well as at a distance.

## POLARIZATION

The principal benefit of the revolution in transport and communications technology has been the enormous expansion in the goods, services, friends and information made available to all those who have access to the new methods of travelling and communicating. But this access has been very unevenly shared. While villages in Africa and India have become more accessible to tourists from wealthy countries, the wealthy are still in little danger of being paid return visits by the villagers they visit. But even within the wealthy countries the benefits have been very unevenly shared. The great increases shown in Table 6.1 in the average number of miles travelled per person per day in the United States and Britain conceal a growing disparity in the individual mobility levels of which the averages are comprised. While on average both countries are becoming more mobile, disparities between major sectors of their populations are very large and growing.

For over thirty years public transport in Britain has been retreating before the advance of private transport. In London the use of public transport reached an all-time peak in 1948 and has been declining ever since. In many less densely populated areas the decline began earlier. In the United States the numbers of people carried by public transit reached a peak in 1926 and has declined steadily ever since, with the exception of the period

during and immediately after the Second World War.(8) Figure 6.3
describes the trends in public and private transport in Britain for
the period 1954 to 1976. Some time around 1958, in terms of passenger kilometres, Britain ceased to be a predominantly public
transport society and became a private one. Figure 6.4, based on
the official forecasts presented to public inquiries in 1979, describes the government's expectations for these trends for the
next twenty-five years.

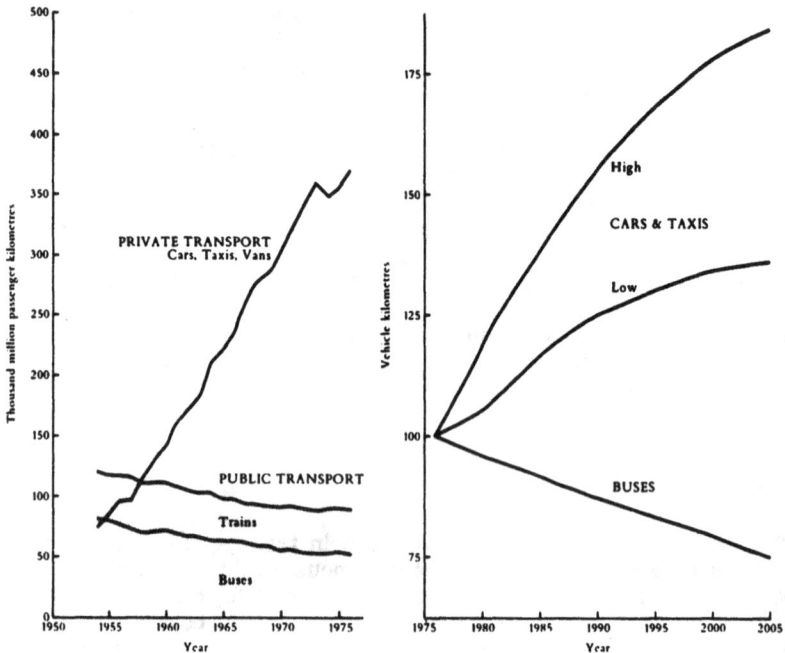

Figure 6.3 *left* Passenger transport
Source: 'Transport Statistics Great Britain 1966-1976', table 10.

Figure 6.4 *right* Forecasts: index of vehicle kilometres (1976 = 100)
Source: 'National Traffic Forecasts: Interim Memorandum, 1978',
        Department of Transport.

The relationship between the trends in public and private transport is clear. The decline in the quantity and quality of public
transport services in Britain since the war has been caused primarily by a massive loss of paying customers who have abandoned
public transport for their newly acquired cars. As a consequence,
overhead costs have had to be shared among a diminishing number of customers, leaving the operators with the choice of increasing fares, or reducing services, or, most commonly, both. As a
result even more customers have been lost.

It is important to distinguish between the causal and consequential aspects of the decline. Public transport used to be better
and cheaper than it now is, and profitable. But even at its best it
could not compete with the car in terms of convenience, comfort
and flexibility. When people became wealthy enough to afford these
advantages, they bought them and were rarely seen on public
transport again. Even if subsidies were to be provided sufficient
to restore public transport to its former glories it is unlikely that
this would do much to stem the flow from public transport to private, let alone reverse it.

Subsidies can, however, reduce the consequential losses of
customers who abandon public transport because it has become too
expensive, and of customers abandoned by public transport when
services are withdrawn. Subsidies can also provide a financial
cushion for the captives of public transport who are obliged to
find the money, somehow, for essential journeys when no alternative mode of travel is available.

The longer the primary losses continue the more expensive it
becomes to prevent the secondary, consequential, losses by maintaining the same service at the same price. Consider the example
of a rural bus that regularly carries fifty passengers. If one of
the fifty passengers were to acquire a car, the remaining forty-
nine passengers could take up the car owner's share of the bus
overhead costs without too much strain. But when the forty-ninth
passenger leaves, the remaining passenger immediately has his
(or more likely her) share of the overhead costs doubled.

The 1975 public transport mileages for the United States given
in Table 6.1 do not convey fully the extent of the country's dependence on the car. A substantial proportion of the public transport mileage consists of commuter traffic. Because of the sparse-
ness of the networks of services, public transport for a great
many Americans is something to which they must drive in their
cars. The figures do, nevertheless, suggest the way in which
the decline in public transport in Britain is likely to accelerate if
British levels of car use continue to approach those prevailing in
the United States now. In 1977 the average level of mobility in
Britain was approximately equal to that in the United States in
1950. Per capita car mileage in Britain in 1977 was slightly lower
than in the United States in 1950 and per capita use of surface
public transport slightly greater. The main difference between
Britain's pattern of travelling in 1977 and the United States's pattern in 1950 is in the mileage travelled by air.

Between 1954 and 1977 the percentage increase in travel by car
in Britain was more than twice that for the United States between
1950 and 1975. But the percentage decrease in the use of public
transport in Britain was less than half that which occurred in the
United States. In the 1970s in Britain the operating losses of most
public transport services increased dramatically, and very many
services are now dependent for their very existence on subsidies.
If Britain were to emulate in the next twenty-five years the
changes that occurred in American transport between 1950 and

1975 the consequences for public transport would be highly predictable.

A report by the Transport and Road Research Laboratory
entitled 'The Effect of Car Ownership on Bus Patronage' (9) contains some very useful evidence for speculating about the magnitude of the consequences. It estimates that in 1976 for every
additional car that took to the roads there was an annual loss to
the bus services of 300 trips. This loss it calls the 'direct effect
of increasing car ownership'. In addition it estimates that there
would be 'second-round' losses caused by the reduction in services and increased fares resulting from the first-round losses,
which would be almost as great. In addition to these effects would
be the longer term 'third-round' losses resulting from people
moving out of the inner areas of cities to outer suburbs where
no public transport services have ever existed.

These losses can be placed in perspective by comparing them
with existing levels of bus usage. In 1977 there were an estimated
7505 million bus trips made.(10) If the present number of cars in
the country were to increase to reach the 26 million envisaged by
the government's forecasters, by 2005 there would be another 12
million cars on the road. If each of these were to be responsible
for the direct loss of 300 bus trips the increase in car ownership
would result in a loss to the bus services of 3600 million trips. If
the second- and third-round effects together were approximately
as great as the direct effect the bus services would have almost
no customers left.

Both Britain's principal political parties are simultaneously
supporters of public transport and promoters of the main cause of
its decline. But a comparison of their policies for dealing with the
plight of public transport reveals a significant difference. The
approach of the Conservative Party is contained in 'The Right
Track: A Paper on Conservative Transport Policy' by Norman
Fowler, the Secretary of State for Transport at the time of writing. 'The Right Track' expresses alarm at the rate of increase in
subsidies for public transport experienced in the 1970s. It es-
pouses the maximum possible choice for the user of transport, and
rejects subsidies for public transport, while conceding that the
phasing out of some of them may take a while; its principal prescription for improving public transport, and it declares itself in
favour of public transport, is increased efficiency and productivity. It rejects 'deliberately anti-motorist policies' declaring that
'the importance of the car and the rights of the motorist should
be recognized in policy-making'. It favours more road building,
with a particular emphasis on bypasses, relief roads, and roads
that will assist economic growth. The central contradiction in the
Conservative policy of maximizing free choice for the user of
transport is that it would, if pursued for long enough, culminate
in a complete absence of choice; if sufficient people choose the
car, the public transport alternative will simply disappear.

The Labour Party is also committed to a great increase in the
number of motorists and is equally vigilant in defence of their

rights. But it is much more eloquent about the plight of public transport users and much more generous in the assistance it offers to them. In the campaign preceding the 1979 general election the then Secretary of State for Transport, William Rodgers, placed considerable emphasis on his party's achievement in almost doubling the amount of public money spent on public transport subsidies in comparison with that spent by the previous Conservative government:

> The Government's own record is plain and unequivocal. In 1973-4, the last year of Mr Heath's Government, less than £350 million was paid in subsidies for public transport and concessionary fares: today a Labour Government is spending well over £600 million. These figures are in real terms and represent a massive shift in resources.

> For the first time it has been recognized that buses as well as trains need substantial public support if fares are not to become prohibitive while public transport services decline.

This is a succinct statement of the choice offered by the major political parties of policies for dealing with the problems of public transport. Under the Conservatives, while services decline fares become prohibitive. Under Labour, while services decline subsidies become astronomical. The financial state of many public transport services, even with large subsidies, is precarious. The Conservatives threaten a drastic reduction of subsidies. The Conservative prescription for many ailing services is sudden death. The Fabian alternative is a slow lingering one.


## SIXTY PER CENT ROYALTY

The persistently repeated fiction that the private car plus a high-speed, high-capacity road network provide what Reyner Banham, quoted in Chapter 3, has called the 'ideal democratic' transport system, is the converse of reality. They provide a system of transport that is inherently undemocratic. Estimates of the percentage of the population ever likely to qualify as car drivers, even given the improbable assumption of universal affluence, suggest that there will always be a minimum of 40 per cent who can never have full participating rights in a car-owning 'democracy'. This 40 per cent includes some people who are incapable of using any form of transport unaided or unaccompanied. But it also includes very many more people who would be capable of the independent use of public transport services if such existed. They include children and teenagers, those becoming infirm through old age, those with defective vision, or a great variety of other minor physical and nervous disorders, those chronically or intermittently under the influence of drugs or alcohol, certain of the mentally handicapped, and all those who have been deprived of the right to drive because of serious infringements of the traffic laws. If we reject the assumption of universal affluence as unrealistic and concede that in the foreseeable future there are also

going to be very many people too poor to buy cars, then it
appears unlikely that those with full participating rights will ever
be in a majority.

A society whose transport system is based on the private car
can never be a true democracy. At its theoretical best it can only
ever be an aristocracy in which 60 per cent of the population can
enjoy a grossly debased version of the kingly privilege of motoring described by Lewis Mumford at the beginning of this chapter.
The rest will be second-class citizens dependent for their mobility
on their feet, taxis, the withered remains of the public transport
system, or the charity of car-owning aristocrats. In the United
States, the most popular 'car-owning democracy' in the world,
there are over 100 million such people. There is reason to suppose
that a great many of them are profoundly alienated by their
second-class status.


## A DIGRESSION ON ALIENATION

Alienation is a much overworked word and carries a heavy burden
of possible connotations, so I will digress briefly to explain the
sense in which I use it. Alienation is a state of mind. It is perhaps an unavoidable state of mind, experienced by everyone in
the contemplation of his own cosmic insignificance. Our relationship to the infinite and eternal is, by its very nature, estranged.
It is not something we can do anything about so we try not to
think about it. But our relations with our fellow men are more
immediate and provide a context within which we have the possibility of being recognized as something more than impersonal
bundles of matter in motion. In a social setting we have at least
the possibility of being wanted, respected, needed, loved - of
being significant. It is only this possibility that makes life bear-
able; and it is denial of this possibility that I call alienation.

It is a condition experienced by Kafka's K. in his struggle
against the inscrutable authority of the Castle; by Orwell's Win-
ston in his fight to retain the few shreds of hope and dignity that
the tyranny of '1984' would deny him; and by Bernard and the
Savage in their contemplation of the meaninglessness of the hedon-
istic Utopia of Huxley's 'Brave New World'. It is experienced by
millions: by the unemployed; by those who do mindless, tedious
jobs; by those who commit 'mindless' acts of vandalism and violence; by those who drop out; by the poor living alongside the
rich; and by all those whose lives are buffeted by an authority
whose scale and complexity are beyond their understanding. It
quite likely is, as Theodore Roszak insists, 'the disease from
which our civilization is dying'.(11)

But for some people alienation does not exist. How can one demonstrate its existence to the sceptical social engineer such as
Stafford Beer who quotes Kelvin's dictum 'Whatever exists, exists
in some quantity and can therefore be measured'?(12) States of
mind cannot be measured. Or what evidence could convince a

doubting behavioural technologist, such as B. F. Skinner, who
recognizes no social reality that cannot be measured as a stimulus
or a response, and who dismisses speculation about the mental
states that mediate between stimuli and responses as 'unscientific'.(13)

If it is a disease, must it not at least have symptoms that can
be measured? This was a line pursued with rather limited success
by Durkheim in his study of suicide. 'Anomic suicide', he reasoned,
was a rejection of life by those who felt rejected by life.(14) The
frequency of this 'response' could therefore be used as an index
of alienation. Unfortunately there is such a diversity of possible
social and economic 'stimuli' that could account for it, he was unable to produce convincing statistical support for the hypothesis.
Others consider crime statistics a useful proxy for alienation.
Crime, they reason, is a rejection of authority by those who feel
rejected by authority. Alcoholism, drug addiction, divorce, mental illness, worker absenteeism, and industrial sabotage are other
possible measures. But the difficulty is that there is an infinite
variety of ways in which people can manifest a state of alienation.
What is probably the most common of all, leading a life of quiet
desperation, is the one most likely to pass completely undetected
by a quantifying behavioural scientist.

I can think of only one symptom that is likely to be common to
all cases of the disease, but it is one that will not be of much use
to the social engineer: as individuals become, and come to feel,
less significant in the eyes of society's planners, their behaviour,
in the eyes of these planners, will become more irrational. Their
behaviour will increasingly be related to alternative ideas of
society that are quite alien to the planner. But this, as I have
already indicated, is not much help; irrationality is as difficult to
measure as alienation.

Alienation is a metaphysical idea. That the material circumstances of people's lives and their relationships with others do
influence their assessment of their own significance can be doubted
by very few. But there can never be a convincing scientific demonstration of this proposition because there are no agreed units
by which self-esteem can be measured. The argument that the
trends in mobility described above are aggravating a state of
mass alienation can only be judged against the unscientific criterion of plausibility.


'MOBILITY LANDSCAPES' AND 'THE KNOWN WORLD'

The graphs depicting centred interaction fields presented earlier
indicate the nature of the change that takes place in the spatial
and temporal dimensions of a society's activities as its mobility
increases. The illustrations have assumed that all members of
society share equally in the increase. Figure 6.5 represents an
attempt to illustrate what happens when some become less mobile
while others become more mobile. It is a highly impressionistic

cross-section through a 'mobility landscape'. No scale is provided
because the interaction fields shown can represent a number of
different levels of aggregation from the individual to the international scale. The interaction fields of the increasingly mobile
spread and overlap, while those of the decreasingly mobile contract.

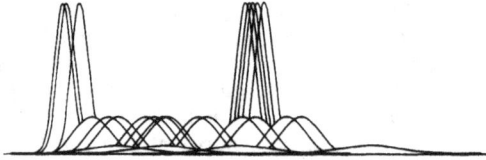

Figure 6.5 Cross-section through a mobility landscape
Source: Adams, J. G. U. (1971), London's Third Airport,
'Geographical Journal', vol. 137, pt 4.

Such changes have important consequences for the way in which
the people affected experience the world. If we define that part
of the world of which an individual might have intimate first-hand
experience as his 'known world', and if we somewhat arbitrarily
delimit this known world by a circle whose radius has a length
equal to one day's strenuous travel, then we can describe what
happens to an individual's known world as improvements in transport technology increase the distance he can cover in one day. If
we make the very generous assumption that a vigorous pedestrian
can move at 8 kilometres an hour and cover 80 kilometres in a
strenuous day, then using the above definition we could describe
the known world of a typical individual in a pedestrian society,
before the beginning of the transport revolution, as the area contained within a circle having a radius of 80 kilometres. This area
is larger than the area assigned to many pedestrian tribal groups
on ethnic maps of Africa. The definition is admittedly rough and
arbitrary but will be seen to be conservative for purposes of the
discussion that follows.
  The bottom circle of Figure 6.6 places a world of this magnitude
in a more familiar setting; it illustrates a circle with a radius of
50 miles (80 kilometres) centred on London's Trafalgar Square.
Ascending from this circle are progressively larger 'known worlds'
whose radii have been increased in the same proportion as the
increase in the speed of transport available. The final circle represents the known world of someone who can travel at the speed
of Concorde. The whole world lies within a day's journey. The
final dot describes what might be called the 'electronically known
world' of all those who are linked to the global telecommunications
network.

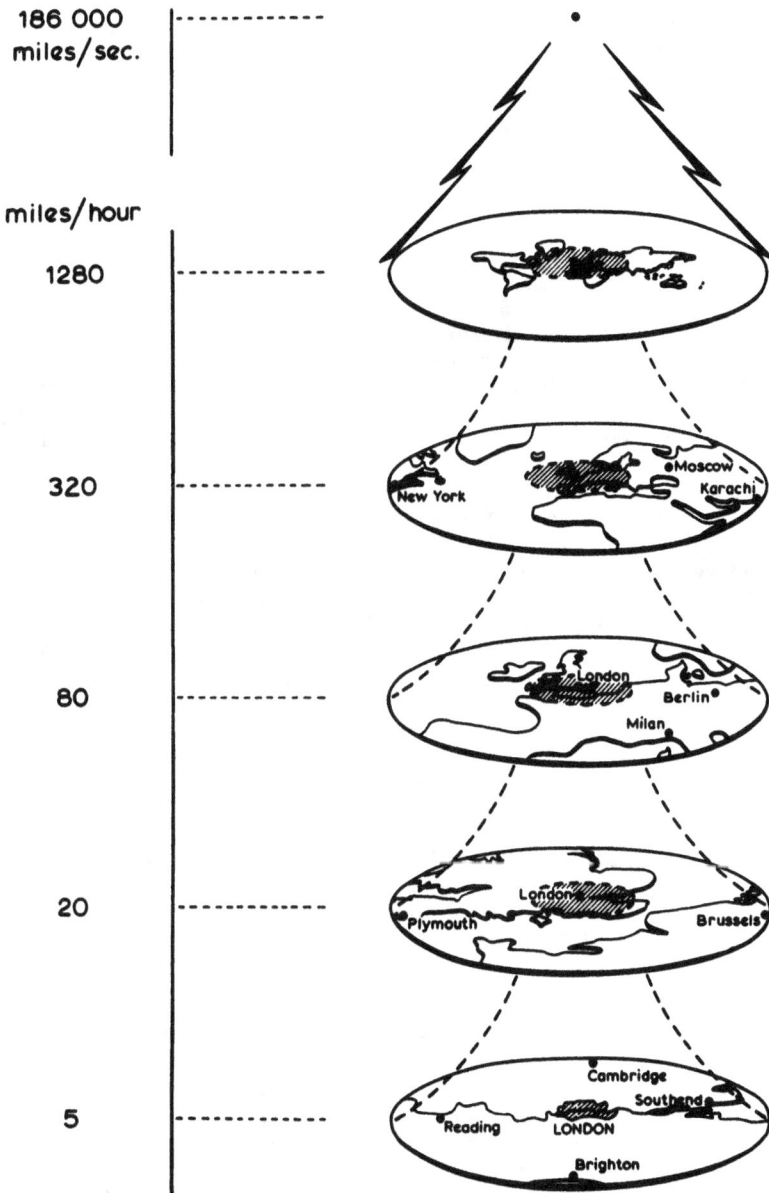

Figure 6.6
Source: 'Environment and Planning', 1972, vol. 4, p. 386

PERCEIVING THE KNOWN WORLD

Although behavioural psychologists and cyberneticians have not
provided any insights into states of mind, and cannot if they
refuse to acknowlege their existence, they have provided some
useful information about human abilities to recognize patterns, to
calculate, to make choices, and generally to cope with the complexity of reality. The psychologists have discovered that in
certain very important respects these abilities are extremely limited, and the cyberneticians have developed, they claim, a number
of useful tools for extending them.

Let us look first at the limitations. Attempts have been made to
estimate the information carrying capacity of the nerve fibres
leading into the brain (15) but, although the estimate is a number
of impressive size, $10^7$ bits/second, relative to the volume of
information that is latent in the world about us, it is insignificant.
The conclusion must be therefore that human information sensors
and processors are highly selective. In the process of visual perception, for example, people separate, somehow, objects of principal interest from their surroundings and view them at an
appropriate level of resolution. Without these selective skills
people would perceive no meaningful patterns - witness the difficulties experienced by someone looking for the first time through
a telescope or microscope. The level of resolution that is selected
is related to the size of the phenomenon being perceived; but
whatever the level selected, a higher level is always possible. This
means that people make sense out of the information impinging on
their senses by ignoring, or filtering out, most of it.

Given man's limitations, and the nature of the perceptual devices employed to compensate for them, we can expect that as the
size of someone's 'known world' increases, there will be a corres-
ponding reduction in the level of resolution at which he 'knows'
the world. For example, few of us, with out modern mobility, now
know a world such as Thoreau knew:

My vicinity affords many good walks; and though for so many
years I have walked almost every day, and sometimes for
several days together, I have not yet exhausted them. An
absolutely new prospect is a great happiness, and I can still
get this any afternoon. Two or three hours' walking will
carry me to as strange a country as I ever expect to see. A
single farmhouse which I have not seen before is sometimes
as good as the dominions of the King of Dahomey.(16)

The detailed knowledge that a hunter will have of a patch of
forest or that a peasant will have of the fields he tills will be quite
different from that of someone driving past on a highway. And the
view from the highway will be quite different from that perceived
from a high-flying airplane. The amount of detail contained in
these views is held relatively constant by a progressive downward
shift in the level of resolution. But what is seen is quite different.
How different was vividly illustrated by an incident during a space
flight in the early 1970s. Although Los Angeles was completely

obscured by smog, a dense, choking smog that made life in Los
Angeles extremely unpleasant, the view of Los Angeles from the
Apollo spacecraft was simply 'beautiful'.

Human scale detail and variety are also lost as the level of resolution is reduced. People are swallowed up by social, national,
and racial stereotypes, by enumeration districts, age-sex pyramids,
population densities, and many other taxonomic devices that social
scientists use for coping with social complexity.

When formalized these processes of selection and shrinking produce what a cybernetician calls 'homomorphic models of reality'.
The construction of such models requires the selection of only
phenomena that can be measured and that are capable of representation mathematically. Once reality has been reduced to a set
of mathematical relationships, it is possible to relate models at
different levels of resolution to each other in a precise and explicit
way. Further, the relationships can be described in a way that
permits information about them to be stored and manipulated outside the head of the individual perceiver; it is this that permits
the cybernetician to deal with a level of complexity far in excess
of that manageable by the unaided human intellect. The growing
demand for the cyberneticians' complexity-managing skills is a
problem considered further in Part III.


THE WIDENING GULF IN COMMUNICATIONS

Crime is endemic in major American cities. Because of an absence
of statistical consistency it is not possible to make precise comparisons between countries of their levels of crime. But measured
by what is perhaps the least ambiguous criminal statistic, murder
and non-negligent manslaughter, the American lead over the
United Kingdom appears to be at least as great as its lead in
mobility. In Northern Ireland in 1976 there were 174 deaths connected with the civil disturbances, including army and police
fatalities, a rate of 11.3 per 100 000 population.(17) In the fifty-
nine cities in the United States with populations over 250 000 the
rate of murder and non-negligent manslaughter in the same year
was almost double, 21.4 per 100 000.(18)

Figure 6.7 describes the trends in Britain for the first three-
quarters of the twentieth century of car ownership, and indictible
offences known to the police, usually considered the most reliable
index of serious crime. There are basically three distinct possible
explanations for a relationship as close and durable as that illustrated by Figure 6.7: the rise in car ownership has been caused
by the rise in crime, the rise in crime has been caused by the rise
in car ownership, or, the association is purely fortuitous. The
first explanation is unlikely, the second and third are both possible.

In its strength and duration the correlation displayed in Figure
6.7 compares favourably with the relationship between car ownership and gross national product commonly used by forecasters for

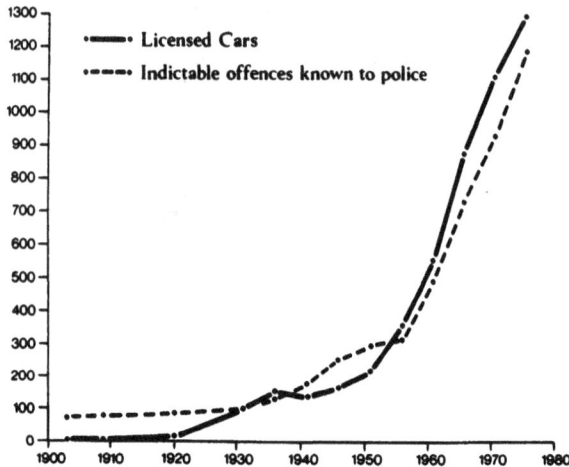

Figure 6.7
Sources: 'Social Trends 1977', HMSO; Carson, W. G. and
Wiles, P. (eds), 'Crime and Delinquency in Britain', 1971,
Figure 1; 'Transport Statistics Great Britain 1966-1976',
Table 44.

predicting car ownership (see Chapters 9 and 10). While it would
be foolish to claim that the trend in car ownership is the exclusive cause of the increase in crime, there are reasons for believing
that the association is not entirely fortuitous. More mobility increases both productivity and opportunity; if it does this
for travelling salesmen, there is no reason why it should not do
it for villains also. Secondly, it fosters anonymity and thereby
encourages the commission of crime by increasing the difficulty
of apprehending offenders; communities in which people are known
to each other tend to be largely self-policing because of the much
greater risk of miscreants being found out.

More importantly, the growing disparities in levels of mobility
foster resentment and undermine the moral consensus that is an
essential precondition for a voluntarily law-abiding community.
Even when they live in close physical proximity to each other the
mobile wealthy and the immobile poor inhabit different known
worlds. If people live in different known worlds they are likely to
develop different group loyalties. Those who inhabit the least
desirable known worlds are likely to rationalize their resentments
and develop ethical codes that encourage a redistribution by theft
of wealth in their favour.

But the principal victims of crimes committed by the immobile
poor are the immobile poor. The arson, robbery and violence characteristic of American urban ghettos suggest that where disparities in opportunity are greatest the resentment of the least well-

off can manifest itself in self-destructive rage. Ghetto residents
are prisoners of their limiting circumstances. The high confined
time-space domes within which they live may be statistical abstractions but they have an unyielding strength. Their occupants are
confined by their lack of mobility in prisons with invisible walls.
They are continually tempted and taunted, in a way that prisoners confined in jails with opaque walls are not, by the freedom and
conspicuous consumption of the affluent. The wealthy can be seen
and heard flying overhead, or driving along motorways through
the ghetto, or on television, enjoying privileges that remain tant-
alizingly out of reach. Prisoners in conventional jails sometimes
vent their resentment and frustration in riots, and so do the
prisoners of the ghetto.

Following the widespread and enormously destructive rioting
that took place in a number of American cities in the summer of
1967, President Johnson appointed a commission to investigate the
circumstances of the riots and to explain why such large numbers
of people were apparently going berserk. The commission's unsur-
prising conclusion was that the riots were race riots and that the
participants were expressing their dissatisfaction with life in their
ghettos. Among the features of ghetto life that most impressed the
commissioners was the fact that 'Poor families in urban areas are
far less mobile than others'. Further they noted that, in the
country with the world's highest levels of car, telephone, and
television ownership, there was a 'widening gulf in communications
between local government and the residents of the erupting
ghettos'.(19)

The root cause of this 'communication gulf' is described in
'Invisible Man' by Ralph Ellison, a black American, as a problem
of perception.(20) White Americans, he insists, simply cannot see
the people who live in black ghettos. 'That invisibility to which I
refer occurs because of a peculiar disposition of the eyes of those
with whom I come into contact. A matter of the construction of
their inner eyes, those eyes with which they look through their
physical eyes upon reality.'

Although Ellison is concerned to argue that white people view
the world through a racial filter that prevents them from seeing
black people, this perceptual problem can only be exacerbated by
the disparity in the levels of resolution at which the fast-moving,
high-flying white American, and the immobile ghetto resident,
view the world. The consequence of this perceptual problem is
alienation, an alienation that Ellison flaunts provocatively:

I can hear you say, 'What a horrible, irresponsible bastard!'
And you're right, I leap to agree with you. I am one of the
most irresponsible beings that ever lived. Irresponsibility is
part of my invisibility; any way you face it, it is a denial.
But to whom should I be responsible, and why should I be,
when you refuse to see me? And wait until I reveal how truly
irresponsible I am.

## CONCLUSION

Modes of transport differ greatly in terms of the numbers of people who have access to them. The most democratic form of transport is a private mode, walking. One of the least democratic modes is a form of public transport available to anyone with the price of a ticket, Concorde. Other modes that are democratic, or potentially so, are bicycles, buses and trains. Private cars, sub-sonic aircraft, both public and private, passenger ships and private yachts are at present accessible to only a tiny fraction of the world's population. Although these 'elitist' modes of transport would become more widely available in an age of universal afflu-ence, for the foreseeable future they will remain the preserve of a privileged minority: there will always be a large proportion of the world's population without the aptitude for driving; the acute problems that most of the major focii of international air travel are having in finding new airport sites suggest that even if a lack of fuel or a lack of money do not prevent aviation becoming a form of travel used regularly by the masses, a lack of airport capacity will; passenger ships have become virtually obsolete as a means of passenger transport, having been transformed into floating holiday resorts, and private yachts and aircraft could not become transporters of significant numbers of people without creating insuperable parking and traffic control problems.

Almost everywhere in the world, governments in consort with the world's principal manufacturing enterprises are devoting most of the resources available for transport to two modes of transport, the car and the airplane, that have no prospect of ever becoming democratic. The result is a world increasingly divided against itself.(21)

## REFERENCES AND NOTES

1  Mumford, L. (1964), 'The Highway and the City', Secker & Warburg, p. 176. Originally published in 'Architectural Record', 1958.
2  Federal Bureau of Investigation (1972), 'Uniform Crime Stat-istics', Table 5. The estimates of the number of cars in China are from the appendix to Chapter 1.
3  This section is an abbreviated and simplified version of argu-ments presented in Adams, J. G. U. (1971), London's Third Airport, 'Geographical Journal', vol. 137, Part 4.
4  Cited in Anderson, J. (1970), 'Space-Time Budgets', Discus-sion Paper 40, London School of Economics Graduate School of Geography.
5  Wilmot, P. (1970), in 'Developing Patterns of Urbanization', P. Cowan (ed.), Oliver & Boyd.
6  Cited in Adams (1971), op. cit., p. 487.
7  'The World Almanac' (1978), p. 423.
8  'Statistical History of the United States from Colonial Times

to 1970', Basic Books, NY, 1976, Series Q241.

9   Oldfield, R. H. (1979), 'The Effect of Car Ownership on Bus Patronage', Transport and Road Research Laboratory, LR872.

10   'Transport Statistics Great Britain 1967-1977', Table 67.

11   Roszak, T. (1970), 'The Making of a Counter Culture', Faber & Faber.

12   Beer, S. (1967), 'Management Science', Aldus Books.

13   Skinner, B. F. (1972), 'Beyond Freedom and Dignity', Jonathan Cape.

14   Durkheim, E. (1963), 'Suicide', Routledge & Kegan Paul.

15   Barlow, H. B. (1968), Sensory Mechanisms, the Reduction of Redundancy, and Intelligence, in 'Cybernetics', C. R. Evans and A. D. J. Robertson (eds), Key Papers, Butterworths.

16   Thoreau, H. D. (1862), Walking, in 'Essays and Other Writings' by Walter Scott.

17   'Social Trends' (1976), HMSO, Table 11.18.

18   'World Almanac' (1978), p. 965.

19   'Report of the National Advisory Commission on Civil Disorders', Bantam Books, 1968.

20   Ellison, R. (1952), 'Invisible Man', Random House.

21   Three books that develop this theme most persuasively are 'Energy and Equity' by Ivan Illich (Calder & Boyars, 1974), 'Personal Mobility and Transport Policy' and 'Transport Realities and Planning Policy', both by Mayer Hillman, Irwin Henderson and Anne Whalley (Political and Economic Planning, 1973 and 1976).

# 7  DEATH

... let us assume that the U.S. authorities had made a [techno-
logical assessment] of the automobile in 1890. Assume also
that this study came up with an accurate estimate that its
use would result eventually in more than 50,000 people a
year being killed and maybe a million injured. It seems clear
that if this study had been persuasive, the automobile would
never have been approved. Of course some now say that it
never should have been. But we would argue that society is
clearly willing to live with this cost, large and horrible as it
is. In Bermuda, which restricts drivers to 20 miles an hour,
there are almost no fatal accidents except with cyclists....
Similar speed limits could be introduced in the United States
if they were wanted, but the majority of Americans apparently
prefer 50,000 deaths a year to such drastic restrictions on
their driving speeds.(1)

This represents a commonly held point of view. Hang-gliding,
mountain climbing and sky-diving are also dangerous, and yet
those who participate in these sports presumably judge the price
worth paying. If the rewards of something outweigh the costs,
then so be it. One might lament the costs but still be prepared to
pay them.
It is notoriously difficult to discern society's judgments on complex issues, but there is some evidence to support the view that
society considers that it is getting good transport value for the
price it is currently paying in life and limb. British Members of
Parliament, who might be presumed to have an interest in assessing the collective will correctly, have judged that society is
willing to pay even more for the benefit of driving faster. In the
discussion in Parliament that preceded the rescinding of the lower
speed limits that had been imposed in the aftermath of the 1973
oil shortage, there was no serious opposition to paying the higher
price associated with higher speed limits, merely a bit of ritual
lamenting:
Mr. John Ellis: Does my right honourable friend agree that
speed is a contributory factor to road accidents and that as
a direct result of his announcement today more people will
be killed and injured on our roads? Are we not guilty of
having a schizophrenic approach to this matter? It is no use
the House or the public throwing up their hands in favour
of safety when such an approach is adopted. Is it not only
fair to say that?

Mr Rodgers (Secretary of State for Transport): Yes, it is
only fair to say that. All of us, including me, suffer from
schizophrenia. We want to save life but we like driving fast.
Although we should all travel slowly, with a red flag in
front of us, people do not choose to do that. We must strike
a balance. It is dangerous in some respects, but that is life.(2)

Thus, in 1977 it was the collective judgment of the Secretary of
State for Transport and the people's representatives in Parliament
that 'striking a balance' involved increasing the number killed and
injured on the roads.

Ideally, of course, everyone would like to get more of what he
wants and pay less, but there are situations in which, to use the
jargon of the economist, the marginal benefits are greater than
the marginal costs. In such cases the economist advises buying
more. The economist's approach to such questions when they involve payment in human lives is discussed in Chapter 10 and
Appendix I. In this chapter an attempt will be made to cast some
light on the question without resorting to the measuring rod of
money.


PREMATURE DEATH

What is being discussed is not death, but premature death. Everyone dies, but there is considerable variation among the nations of
the world in the things from which, and the ages at which, people
die. In the wealthy nations with high levels of road traffic, traffic accidents are a more important cause of death, both relatively
and absolutely, than in poor countries with little road traffic. The
transport systems of the wealthy nations have also been identified
as an important indirect cause of illness and death. In 'The Western Way of Death', Carruthers argues that the combination of the
stress of driving and the lack of exercise endemic in societies in
which most mobility is mechanically assisted, is an important cause
of heart disease:

Stress increases the amount of stress hormones, adrenaline
and noradrenaline, made by the body's sympathetic nervous
system. These stress hormones raise the level of free active
fat in the blood to prepare the body for physical exertion
which, in the modern urban environment, seldom comes. As
a result, the now redundant free fatty acids are laid down
in the walls of the blood vessels as natural fat and choles-
terol. When the coronary arteries have been narrowed by a
critical amount, a final stressful episode makes the blood
supply to the heart insufficient for its needs, often due to
blocking of one of the heart's arteries by the formation of a
blood clot on the roughened vessel wall. The heart then
either rapidly stops beating for lack of blood to its muscular
walls, or suffers painful damage in the part supplied by the
blocked artery.(3)

It is often maintained that the Western way, while not enjoyable, is preferable to the way associated with most other points of the compass. There is a positive relationship between the degree to which a country's transport system depends on the car, and the life expectancy of that country's inhabitants. In general, people in the richest countries tend to own more cars and live longer than people in the poorest countries. There is a positive correlation also between traffic accident mortality rates and life expectancy because the increased risk of being killed in a road accident in a rich country is more than offset by the decreased risk of dying from the multitude of diseases that still afflict the poor. Hence, it is sometimes argued, a high road traffic mortality rate ought to be worn proudly as a badge of 'development'.

Figure 7.1 suggests that the foregoing argument is somewhat simplistic. There is no necessary connection at all between a country's level of car ownership and the life expectancy of its inhabitants. Certainly there are many countries with few cars and high life expectancy levels. If one considers only the countries with life expectancy levels above 65, the distribution shown in Figure 7.1 offers no support for the idea that cars are necessary for a long and healthy life.

Figure 7.2 shows the relationship between car ownership levels and traffic accident mortality rates. It is a fan-shaped distribution showing that there is a wide range in the mortality rates of countries having similar levels of car ownership. The bottom boundary of this distribution is of interest because it indicates the best that has been accomplished in terms of safety for the present range of car ownership levels. Two alternatives are shown for the United

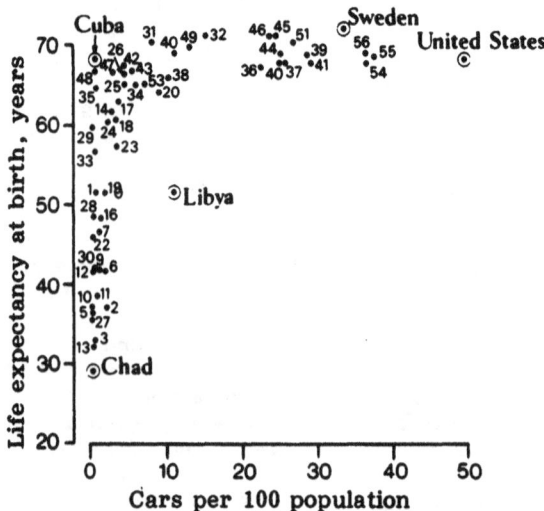

Figure 7.1 Life expectancy (male) and car ownership
Source:  Table 1.2 and 'World Almanac 1978'.

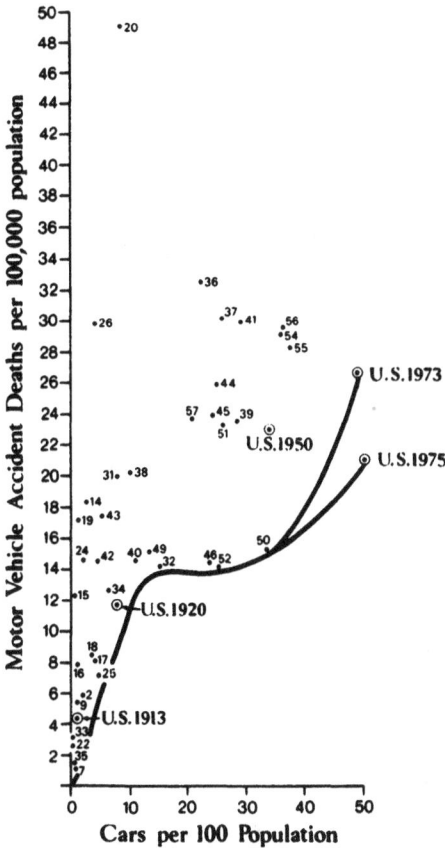

Figure 7.2 Car ownership and traffic accident mortality rates
Sources: Table 1.2 and 'UN Demographic Yearbook 1975'.

Key to Figures 7.1 and 7.2

| | | | |
|---|---|---|---|
| 1 | Algeria | 14 | Costa Rica |
| 2 | Angola | 15 | Cuba |
| 3 | Central African Rep. | 16 | Guatemala |
| 4 | Chad | 17 | Jamaica |
| 5 | Ethiopia | 18 | Mexico |
| 6 | Ivory Coast | 19 | Nicaragua |
| 7 | Kenya | 20 | Trinidad |
| 8 | Libya | 21 | United States |
| 9 | Mozambique | 22 | Bolivia |
| 10 | Nigeria | 23 | Brazil |
| 11 | Senegal | 24 | Chile |
| 12 | Sierra Leone | 25 | Uruguay |
| 13 | Upper Volta | 26 | Venezuela |

| | | | |
|---|---|---|---|
| 27 | Bangladesh | 43 | Hungary |
| 28 | Burma | 44 | Italy |
| 29 | China | 45 | Netherlands |
| 30 | India | 46 | Norway |
| 31 | Israel | 47 | Poland |
| 32 | Japan | 48 | Romania |
| 33 | Philippines | 49 | Spain |
| 34 | Singapore | 50 | Sweden |
| 35 | Sri Lanka | 51 | Switzerland |
| 36 | Austria | 52 | United Kingdom |
| 37 | Belgium | 53 | Yugoslavia |
| 38 | Czechoslovakia | 54 | Australia |
| 39 | France | 55 | New Zealand |
| 40 | East Germany | 56 | Canada |
| 41 | West Germany | 57 | Finland |
| 42 | Greece | | |

States because the permanence of the drop in the death rate associated with the energy crisis speed limits remains to be demonstrated. Figure 7.2 does not preclude the possibility of a country climbing the ladder of car ownership and achieving lower mortality rates, but it offers an indication of what it might reasonably expect in the way of increased traffic fatalities if it were to succeed in numbering itself among the 'safer' countries.

What distinguishes death in traffic accidents from death by most other causes, with the exception of the causes of infant mortality, is its excessive prematurity. In a survey of causes of death in twenty-seven countries in Europe, North America and Oceania (i.e. 'developed' counties) accidents were overwhelmingly the leading cause of death for all ages from 1 to 44;(4) traffic accidents accounted for between 30 per cent of all accidental deaths in Bulgaria, with 11.3 cars per 100 inhabitants, and 56 per cent in Australia, with 36.9 cars per 100. Traffic is predominantly a killer of people in the prime of life. The traffic accident mortality rate typically reaches a peak in the late teens and declines thereafter until it experiences a small rise again after age 65. This distribution of accidents by age appears to be confined to motor vehicle accidents. The non-motor-vehicle accident rate increases with age in a way similar to most other death rates.(5) In a number of countries traffic accidents account for more than half of all male deaths between 18 and 24.(6)

Because there are no agreed units for measuring collective grief it is impossible to offer society any firm statistical foundation upon which to base its judgment about the seriousness of traffic accidents. Compared to the major killers, such as cancers and heart diseases, they might be thought relatively unimportant. In 1972 in the United States ischaemic heart disease, the leading cause of death, was responsible for 684 424 fatalities (35 per cent of all fatalities) while traffic accidents accounted for only 56 278 (2.9 per cent). Thus judged by crude death rates ischaemic heart disease might be considered a problem twelve times more serious than road accidents. But the average age at which people died from ischaemic heart disease was 73.5 while the average age at which people died in road accidents was 35.5.

Although there is almost universal agreement that death is inevitable, there is much less agreement about the fit and proper time at which the inevitable ought to happen. Figure 7.3 represents an attempt to view the prematurity of road accidents from the perspective of the biblical allotment of three score years and ten. From this perspective not all deaths are equal. Those who manage to exceed the biblical allotment by a wide margin are often cheered to the finishing line. Those who fall a long way short are usually mourned for what might have been. The idea that there is 'a time to be born and a time to die' can moderate the grief associated with a death that is judged 'timely'. Multiplying the number of years by which they fell short of the biblical datum gives the number of person-years of which society has been 'cheated' by the particular cause of death. If a disease has a positive person-year balance with respect to the datum it might be considered a 'natural' cause of death.

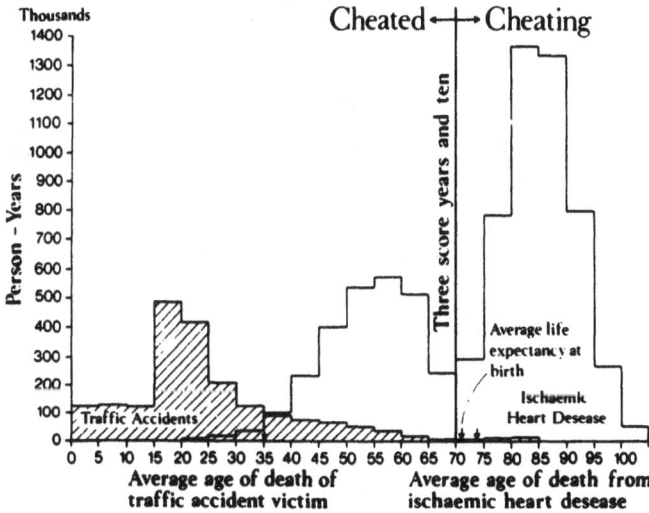

Figure 7.3 Cheating death and cheated by death: US 1972
(Number of deaths in each age group x number of years above or below 70)
Source: 'Vital Statistics of the US 1972', part A, US Department of Health, Education and Welfare, 1976, Table 1.26.

Figure 7.3 shows that road accidents are overwhelmingly a cheating cause of death. Ischaemic heart disease appears to be both a natural and a cheating cause. The relative importance of the contributions of rich diet, stress and lack of exercise to heart disease is a matter of medical controversy. But if a significant share of the premature deaths caused by heart disease can be attributed to the stress and lack of exercise associated with mass car ownership, then the United States's system of transport emerges as a strong contender for the title Public Killer Number One.

A judgment about whether this price is worth paying might also be usefully informed by a comparison with certain other causes of premature death. War, murder and suicide are judged by society, certainly by its information media, to be particularly abhorrent, tragic and senseless because they involve the termination of life both prematurely and deliberately. During the Second World War Britain's enemies deliberately killed 244 723 British soldiers in battle and another 60 595 civilians by bombs and rockets, giving a Second World War total of 305 318.(7) Between 1926 and 1976 331 214 people were killed, inadvertently, on the roads in Britain. (8) The United States has far more cars but is also on the bottom boundary of Figure 7.2; the number killed on the roads between 1913 and 1976, 2 126 712 (9) is more than three times the number of Americans killed in all the wars in which the United States has ever participated - Vietnam War, Korean War, the Second World War, the First World War, Spanish American War, Civil War (both sides), Mexican War, War of 1812 and War of Independence.(10)

Traffic accidents also claim more lives than murder and suicide, the civilian modes of deliberate premature death. In Britain the traffic accident death rate typically exceeds the murder rate by a factor of 14, and the suicide rate by a factor of 2. In the United States the murder rate and suicide rate together roughly equalled the traffic accident death rate in the mid-1970s. The comparison of lives lost inadvertently with the numbers killed deliberately does not of course demonstrate that the number killed in accidents is too high, but it does place in perspective a price that is, as the parliamentary exchange quoted above shows, deliberately paid.

## CONCERN

In all countries with large numbers of cars there is great concern about the accident problem. But the concern is invariably qualified by a concern of equal or greater strength to promote automotive mobility. The following passage comes from a paper presented in 1973 to the First International Conference on Driver Behaviour by B. J. Campbell, the director of the University of North Carolina Highway Safety Research Centre:

> In the USA at least, neither the economy nor the quality of life can willingly tolerate a situation in which large segments of the driving population are denied the opportunity to drive.... Society has an enormous investment in its highway system to achieve mobility. Accidents are a failure of this mobility system, and the key issue is whether accident reducing programs increase or decrease mobility. A program which increases safety *and* increases mobility optimizes the

system. The Intersate Highway System is an example of a
system that increases *both* safety and mobility. On the other
hand, a program which reduces accidents by restricting
mobility or eliminating driving by certain people is mobility
reducing, and the annoyance of reduced mobility is compounded by knowledge of the poor ability to predict who
will have an accident.(11)

The literature on road accidents and road safety is dominated
by the concern to find methods for reducing accidents without
reducing mobility. Such concern might be likened to the tobacco
industry's concern about lung cancer. The industry is in the
vanguard of the search for a safer cigarette but seems less than
enthusiastic about solutions to the cancer problem that involve a
reduction in smoking. In both cases the behaviour that causes the
fatalities is considered desirable in itself. The search for non-
mobility-reducing methods for reducing accidents has been going
on for most of the twentieth century and has turned up remarkably little. Two articles by the late R. J. Smeed suggest that the
reason is obvious.(12) The number of accidents varies in a highly
persistent way with the volume of traffic. Smeed examined the
relationship between the level of motorization and the level of
road accident deaths for a large number of countries over a period of thirty years and reached the following conclusion:

The general tendency shown in the 1938 data has, to a
large extent, remained unchanged in the last 30 years. The
formula suggests that doubling the number of motor vehicles
in a country with a given population resulted - on average -
in an increase of fatalities of 26%. The average increase in
casualties would, however, be expected to be greater than
this because the less serious accidents have tended to increase at a greater rate than the more serious ones.

Although these tendencies may well continue, it is not
possible to predict with certainty that they will do so. There
are definite indications that some accident risks are increasing as roads become more crowded and this suggests that
more active steps may be necessary to reduce accident frequencies.

Neither the tobacco industry nor the transport policy-makers have
made much progress towards solving their respective safety problems because both are committed to seeking solutions that involve
increasing the population at risk.

In 1970 the accident problem in the United States was summarized by Eames, Lee and Fell of the United States National Highway
Safety Bureau: 'Undeniably, the incidence of highway accidents
has reached epidemic proportions. Though much is being done to
help combat the problem, we are not close to satisfactory solutions.
(13)

Figures 7.4 and 7.5 illustrate the progress of this epidemic. The
only index that registers any improvement since 1970 is the death
rate, and this achievement, because of its close association with
the introduction in 1974 of the 'energy crisis' 55 m.p.h. speed

limit, must be primarily credited to the attempt to save oil, not
lives. Since the reduction in fatalities that has occurred is the
result of an oil-saving measure that does not save much oil, and
since it is also associated with a restriction of mobility of the sort
which in Britain was abandoned for lack of political support, it is
an improvement whose long-term prospects might be described as
precarious.

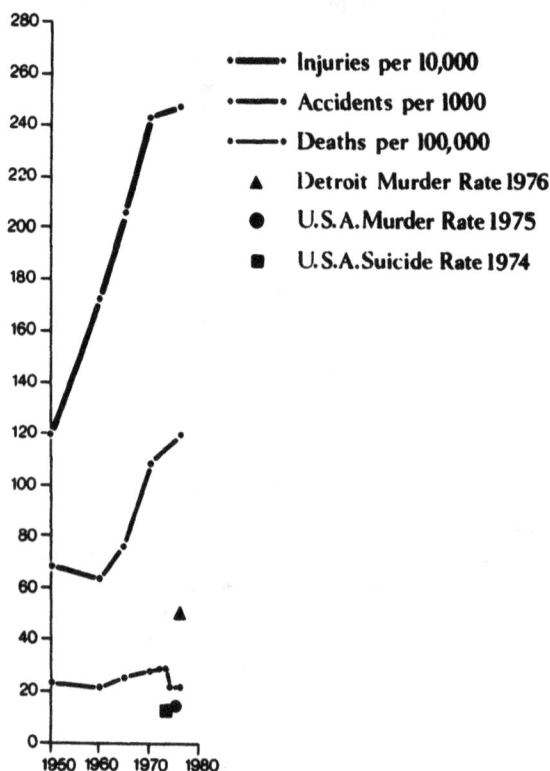

Figure 7.4 Road accidents USA
Sources: 'UN Demographic Yearbook 1975'; 'Statistical Abstract
          of US 1977'.

## A DIGRESSION ON INTERPRETING ACCIDENTS STATISTICS

There is an interesting technical problem associated with interpreting data such as those displayed in Figures 7.2 and 7.3. As
noted by Smeed, the number of non-fatal accidents appears to be
increasing more rapidly than the number of fatal accidents. In the
United States in 1950 there were fifty-two injuries recorded for

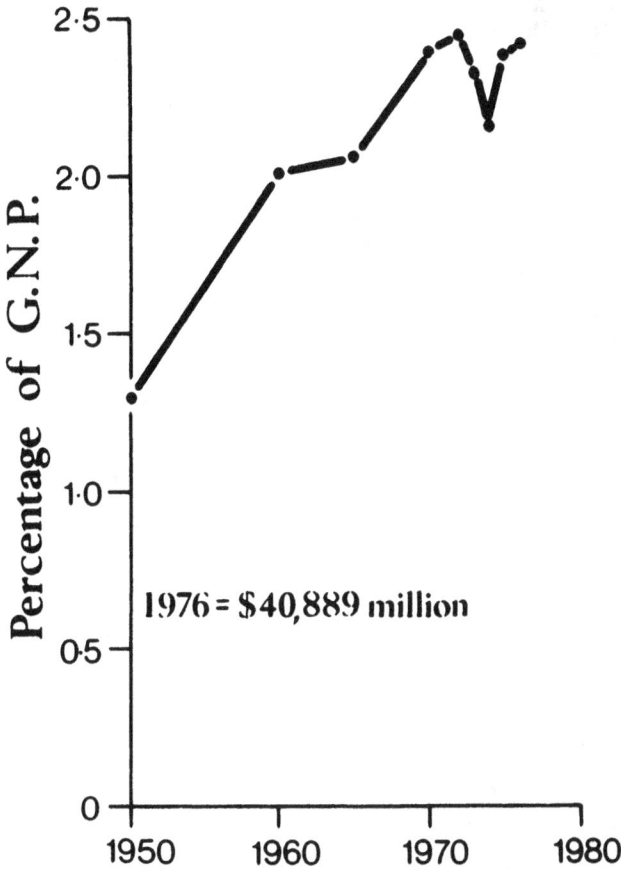

Figure 7.5 The cost of road accidents USA
Source: 'Statistical Abstract of the US 1977'.

every death and by 1976 the figure had increased to 112. A similar trend is apparent in the British statistics; in 1926 there were
twenty-seven injuries recorded for each death and in 1976 there
were fifty-one. In 1937, the first year for which 'serious' injuries
are recorded separately in Britain, there were eight serious
injuries for each death and by 1976 the number had risen to twelve.
This phenomenon may, in part, be explained by changes in rec-
ording practices. The increased availability of 'free' medical treatment could be expected to attract more people to medical services
that keep official records of accidents. Thus part of the increase
in injuries in recent years might be accounted for by an increase
in the proportion of injuries that are recorded. The increase in
the rewards of accident litigation are also an incentive to people
to call attention to the consequences of accidents of which they

were innocent victims, and it is easier to feign injury than death.
For these reasons there is a tendency to consider the injury statistics inflated and to treat the death statistics as the only reliable
guide to the state of the road accident problem.

But however certain and unambiguous death might seem to be,
the road accident fatality statistics consistently, and increasingly,
understate the amount of premature death caused by traffic accidents. Figure 7.6, taken from Smeed, shows why. The fatality
statistics are highly sensitive to the definition of a fatal accident.
The Economic Commission for Europe defines a road traffic fatality as an accident victim who dies within thirty days of the
accident. It is clearly pointless to compare figures not based upon
the same definition, and the variable quality of international statistics suggests that Figure 7.2 must be interpreted with caution.
(14) But even where a consistent definition is employed, the nature of what is embraced by that definition can change over time.
Improved casualty facilities and intensive care units have greatly
improved the chances that an accident victim will survive beyond
the officially recognized fatality period. Thus some lives that are
clearly ended prematurely by traffic accidents will be recorded as
injuries and not fatalities.

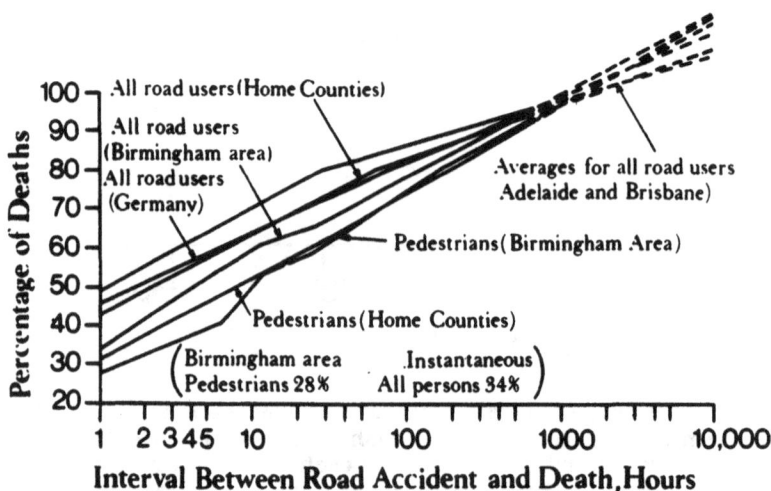

Figure 7.6
Source:  Smeed, R. J. and Jeffcoate, G. O. (1970), Effects of
         Changes in Motorization in Various Countries on the
         Number of Road Fatalities, 'TEC', vol. 12, no. 3,
         pp. 150-1.

There can be no firm evidence on the magnitude of the statistical distortion caused by this problem because the longer an
accident victim survives, the more difficult and arbitrary becomes

the decision whether to attribute death to a complication resulting from his injury, or to the injury itself. The lines on Smeed's graph stopped at thirty days. A projection of the strongly linear trends beyond this point suggest how serious the underestimation of accident-related premature death might be. The fact that the lines rise above 100 per cent simply indicates the arbitrary basis of the definition of a fatal injury on which the original graph was based. Traffic accidents can cause anything from instantaneous death at one extreme to a slight reduction of the victim's allotted span at the other. Most official fatality statistics that record only the first 720 hours of this range are bound to understate the severity of the fatal accident problem.

## A POST-MORTEM ON ROAD SAFETY PROGRAMMES

Explanations for the failure of road safety programmes can be found by examining the three major objectives of the programmes: safer roads, safer cars, and safer drivers.

*Safer Roads*
Road building is a very big business. Between 1970 and 1976 £4.8 billion (1975 prices) (15) was spent on the construction of roads in Britain. In the United States in 1975 construction contracts worth $7.2 billion were awarded for state and federal highway projects.(16) Those who build roads have an obvious interest in seeing the business remain big. Barbara Castle, former British Minister of Transport, has described the road interests as 'the most vociferous lobby in this country'.(17)

The financial power and political influence of this lobby have been extensively documented.(18, 19, 20, 21, 22) As with all successful lobbies, its success depends on its ability to persuade the public that it is serving the public interest. The two products that the road construction industry offers society as justification for its activities are mobility and safety. Undeterred by evidence to the contrary, the industry assiduously cultivates the myth that it provides both. The foundation of the myth that new roads - especially restricted-access divided highways - save lives, is the fact that such roads tend to have lower accident rates per vehicle kilometre. But it is highly misleading to use such a fact in isolation. If the tobacco industry succeeds through the combined efforts of its research and sales staffs in halving the cancer risk per cigarette smoked while more than doubling the number smoked, it does not deserve any credit for reducing cancer. A similar argument fits the activities of the road lobby.

Since the war traffic in most countries has increased at a much faster rate than the capacity of the road system, in spite of the impressive amounts of money spent on new roads. Figure 7.7 shows that in Britain between 1948 and 1976 at no time did the addition of new road capacity catch up with the growth in the number of cars, nor was there any realistic prospect in most

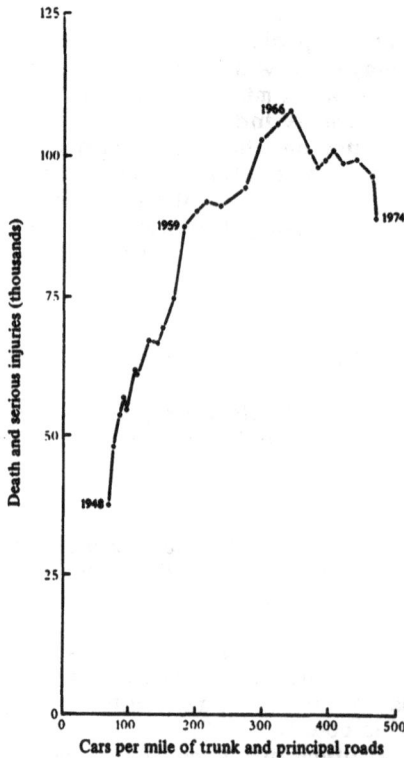

Figure 7.7 Road accidents and road capacity
Source: 'Transport Statistics Great Britain, 1964-74', HMSO.

countries that it could. Throughout this period investment in new capacity was concentrated on those parts of the network that were most congested, that is, those parts where investment, by relieving bottlenecks, would most effectively facilitate the growth of traffic throughout the whole of the road network, most of which remained unimproved. Throughout this period there was in Britain a more than fourfold increase in traffic, and an almost threefold increase in the number of deaths and serious injuries. The decline of the accident graph since 1966 cannot be attributed to the rate of of road building suddenly overtaking the growth rate of car ownership, because it did not. The two sharp drops on the graph are associated with the introduction of the breathalyser and the 'energy crisis'. The first reduced the accident rate per vehicle kilometre, and the second reduced both the accident rate and the volume of traffic. Since the number of accidents is the product of the accident rate and the volume of traffic, the gains in safety made since 1966 will be squandered by current traffic promotion

policies unless - and it would be an achievement without precedent - new safety measures can be devised that will reduce the
accident rate faster than traffic increases.

Of the money spent on new road construction in Britain during
the 1970s almost exactly half has been spent on the Strategic Road
Network, the system of high-speed, high-capacity roads designed
to link the major centres of population. In the United States the
single largest transportation project undertaken since the Second
World War is the 42 550 mile (68 000 kilometre) Interstate Highway
System. The American project, the largest public works project
ever undertaken by man,(23) was originally estimated to cost $26
billion when authorized by Congress in 1956. By 1973 over $40
billion had been spent and the United States Department of Transportation declared in 1975, 'The date at which the Interstate
System can feasibly be completed has continued to recede as the
costs of highway construction have escalated.'(24) It is the construction of roads such as these upon which the financial health
of the road construction industry rests. And it is roads such as
these, with their median strips, wide shoulders, gentle curves
and absence of cyclists and pedestrians, which are promoted, by
safety experts such as Campbell quoted above, on the basis that
they increase safety.

In 1973 the United States had about twice as many cars per
capita as Britain and four and a half times as many kilometres of
road per capita. About 20 per cent of all traffic flowed on the
'safest' limited-access roads in the United States compared to
about 13 per cent in Britain. The per capita road death rate in
the United States in 1973 was 79 per cent higher than in Britain.
Although the statistically rough nature of these comparisons pre-
cludes basing any firm conclusion upon them, they do suggest
that building 'safer' roads has not made a very impressive contribution to safety in either country. What can be said about the
contribution of road building to safety on the basis of the existing statistical evidence has been neatly summarized by Wilde:

It is often stated that some types of roads are safer than
others. Supposedly as proof of this, we are being told that
modern four-lane highways have a lower fatal accident rate
per million kilometres driven than two-lane highways and
secondary roads. However, on four-lane highways traffic
moves considerably faster than on other road types, which
means that the difference in the fatality rate per million
man hours behind the wheel is much smaller. Moreover,
highways that allow high speeds make driving more attractive and convenient relative to other means of transport. It
has been calculated in Canada that per passenger mile,
travelling by train is approximately thirty times safer than
travelling by car. Thus, it would seem more adequate to
say that modern highways are safer per mile, about as safe
per hour of driving, and more dangerous per capita.(25)

It is important to decide whether one prefers miles or people as
the denominator for one's safety index. The former is favoured by

the advocates of more road construction and the latter is preferred by their opponents. In the United States between 1921 and 1973 the death rate per 100 million vehicle miles has dropped from 25.3 to 4.2, while the death rate per 100 000 people increased from 12.8 to 26.5. In 1973 in terms of deaths per vehicle mile the United States was the safest country in the world,(26) but in terms of deaths per 100 000 people, it was, as Figure 7.2 shows, one of the most dangerous.

*Safer Cars*
The design of cars emerged as a popular safety issue in the late 1960s in the United States following the publication of Ralph Nader's book 'Unsafe at Any Speed'. The National Highway and Traffic Administration was formed with powers to set vehicle safety standards. It required a number of design changes such as padded instrument panels, tougher windscreens, energy absorbing steering columns and seat belts whose aim was to improve the chances of a car's occupants surviving a crash. Although it also required design changes aimed at preventing crashes, such as dual brake systems, the emphasis in its regulations was on protecting a car's occupants during a crash.

The most comprehensive study of the effectiveness of these measures is Peltzman's The Effects of Automobile Safety Regulation published in the 'Journal of Political Economy' in 1975. His conclusion: 'The one result of this study that can be put forward most confidently is that auto safety regulation has not affected the highway death rate.'(27)

Fatal accidents are caused, or prevented, by a number of factors, of which vehicle safety is but one. If changes occur in any of the other factors at about the same time that government vehicle safety regulations come into effect, it can be extremely difficult to attribute responsibility for changes in the accident rate. Peltzman's study examined variations in road death rates over time (time series evidence) and space (cross-section evidence) in order to isolate the factors that contributed most to a statistical explanation of the road death rate. These factors turned out to be income, the cost of accidents, consumption of alcohol, driving speed, driver age, and a catch-all factor called 'secular trend' (trend over time). These factors were incorporated in a mathematical model that fits the time series data impressively well. Over 99 per cent of the variation in the death rate between 1947 and 1965 is 'explained' by the model. The model was used to predict what the death rate would have been in 1972 without the vehicle safety legislation. This prediction was very close to the death rate that was achieved with the legislation. The actual death rate in 1972, with the safety regulations in effect, instead of displaying the 20 per cent reduction anticipated in the safety literature displayed a 5 per cent increase above the level predicted, thus leading Peltzman to his conclusion that the regulations had been of no effect.

There is an important reservation to be attached to these results and the conclusion that Peltzman draws from them. His time series

model was used to predict the death rate per vehicle mile, and
this rate did in fact fall after the introduction of the regulations -
by 20 per cent between 1965 and 1972. But this fall represented
a continuation of a 'secular trend' of long standing. Although this
trend is one of the variables in his model he offers no explanation
for it. It could be argued that it represents the effect of a large
number of safety measures applied over a long period of time:
safer roads, safety propaganda, speed limits, more experienced
drivers and pedestrians, better brakes, tyres, suspension,
steering, etc. Peltzman interprets the fact that this downward
trend did not become steeper after the regulations were implemented as evidence that the regulations had no effect. But it
could mean that they had no more effect than previous safety
measures.

However, while there is room for dispute about the contribution
of the vehicle safety regulations to the decline in the death rate
per vehicle mile, there can be no dispute about its contribution to
the decline of the death rate per capita, because this rate increased by 6 per cent.

Peltzman offers a peculiarly economic explanation for the failure
of vehicle safety regulations to make an impact on the trend of the
road death statistics. He suggests that the death rate of a society
at any particular time represents the 'equilibrium price', paid in
lives, of a given level of 'driving intensity', that is, the price
that society is prepared to pay for a given level of mobility. If a
government attempts to impose, by regulation, safer vehicles than
society wants, then society will restore the equilibrium by driving
more dangerously. He notes that very few people purchased voluntarily the safety features that the regulations obliged them to
buy. He also notes that a shift occurred in the accident burden
from car occupants to non-car occupants. A possible explanation
of this shift is that drivers, persuaded that their cars were safer
to have crashes in, became less cautious and, as a consequence,
hit more cyclists and pedestrians. Such an effect, he suggests,
would be 'consistent with optimal driver response to an exogenous
reduction of the expected loss from an accident'.

Peltzman's 'equilibrium' is the economist's version of the 'balance'
that Transport Secretary Rodgers was attempting to restore by
raising the speed limit in Britain. To call 'optimal' the driver response that restored the balance upset by the safety regulation,
suggests that the only sure guide to the number of road accidents
that a society wants is the number that society, in an unregulated
state, actually gets.

*Safer Drivers*
The view that attempts to make roads and cars safer will be frustrated by drivers driving more dangerously is both a theory of
accident causation and a reaffirmation of an article of faith of
laissez-faire economics. The ramifications of this view have been
explored by Wilde in a paper in which he presents the intriguing
idea that drivers have a 'risk thermostat'.(28) He reviews a

persuasive amount of evidence to support his hypothesis that
attempts to alter the setting of the thermostat by government fiat
are virtually certain to fail. He cuts through the forest of factors
causing road accidents and leaves only one tree standing: 'There
is basically only one independent variable which controls the number of accidents and this is risk tolerance.'

Although this explanation of the failure of safety programmes is
essentially the same as Peltzman's, Wilde's approach to the accident issue is quite different. Peltzman opposes 'consumerist'
meddling with market forces and appears to oppose any government interference with the free market in risky driving on the
grounds that it can only hinder the work of Adam Smith's Invisible Hand. He treats the settings of individual thermostats as
optimal and immutable, at least by government action. Wilde, on
the other hand, is in active pursuit of policy measures that will
reduce the settings of the thermostats.

He proposes four types of measures that might alter the level of
tolerated risk: reductions in the rewards of risk - such as paying
truck and taxi drivers by the hour instead of by the trip or kilometre, reductions in the penalties for caution - such as flexible
working hours to reduce the penalty for being late to work;
increases in the costs of risk - such as penalties for not wearing
seat belts; and increases in the rewards for caution - such as
greater reductions in the cost of insurance for accident-free
drivers. His approach is much more constructive than Peltzman's
in the sense that it attempts to do something about the accident
problem, but his proposals are all vulnerable to the laissez-faire
economist's objection that they are unlikely to be effective.
Attempts to regulate the method of payment for truck and taxi
drivers, for example, would be virtually impossible to apply to
self-employed drivers and would put regulated transport concerns
at a very serious competitive disadvantage. Mandatory flexible
hours would wreak havoc with a whole host of commercial activities
that depend on punctual collaboration, and would have no effect
on the driving of someone who wanted to get to work early in
order to get home early. Interference with the rate-setting practices of insurance companies in order to have the riskier drivers
subsidize the safer drivers sounds promising, but Wilde himself
cites evidence that casts doubt on the efficacy of such a carrot
and stick arrangement. An experiment was conducted in California in which 15 000 drivers who had been free of driving demerit
points for one year were rewarded by being exempted from the
requirements to renew their driving licences and submit to an
examination on the highway code. The result was that in the following years the rewarded drivers had between 14 per cent and
46 per cent more accidents than the unrewarded control group.
And finally, Peltzman's evidence and Wilde's own risk compensation
hypothesis both suggest that if the mandatory wearing of seatbelts
were to make crashing safer, drivers would compensate by having
more crashes.

## SOCIETY'S JUDGMENT

The evidence reviewed so far suggests that the road accident problem is a highly intractable one. Safety legislation represents a response on the part of the legislators to what appears to be a real demand on the part of society for more safety, and yet society's response to this legislation has persistently nullified its intended effect. This response has been widely interpreted as reflecting 'society's judgment' that society is prepared to live - and die - with the problem.

Accidents, by definition, are events that no member of society intends. In the case of car accidents an extremely small proportion can be attributed to unpreventable mechanical failure or unanticipatable road hazards. Most are caused by lapses of concentration, errors of judgment, or both. Both are extremely difficult to detect before the event and glaringly obvious after it. They occur in a highly, but not totally, random manner. It is possible to identify groups within the driving population that consistently have more than their fair share of accidents. Young men who drink, for example, have far more accidents than middle-aged women who abstain. People with previous histories of accidents or traffic law violations can also be relied upon, as a group, to persist in their bad habits.

But statistical knowledge of this sort is not very helpful in the fight against accidents. Everyone is young at some time in his life, about half the population is male, and about three-quarters of the adult population of the United States drinks (29) (about 80 per cent in Canada).(30) Thus it is most unlikely that a political majority could ever be found to support the banning of the most accident-prone groups of drivers. And in any event, Campbell's study in North Carolina suggests that banning people with previous records of accident proneness would not make a significant contribution to the reduction of accidents. The worst 1.3 per cent of North Carolina drivers had two or more accidents in the two years prior to the study, but banning them would not have prevented 96.8 per cent of the accidents that occurred in the following two years. He concludes that, while it is possible to identify accident-prone groups with two or three times their share of accidents, most of the accident problem is caused by 'normal but fallible drivers'.

Campbell argues that attempts to reduce accidents by restricting or eliminating driving by the accident prone would be resented because of our 'poor ability to predict who will have an accident'. Although it is not possible to predict how many accidents an individual will have, or how serious they will be, it is not very difficult in the United States to predict who will have an accident. Almost everyone will be involved in a road accident at some time during his life. If the 1973 rate of one road accident for every eight persons were to remain constant for the next seventy-three years - the life expectancy of a new-born American - then the typical new American could expect to be involved in about nine road accidents before he died. He could also expect to be injured once or twice, and he would stand more than one chance in fifty of ending his life in one of these accidents.

It is the normality of lapses of concentration and errors of judgment that explains the intractability of the accident problem. The example of drunken driving suggests that it is likely to remain intractable. Driving while impaired is frequently identified as the single most important cause of road accidents.(31, 32) Although there are other contenders for this honour there is certainly much statistical justification for calling it a major cause. Most legislators, and the people who vote for them, are fallible drivers who drink. In most countries in the world the consumption of alcohol is increasing. In Ontario the consumption of alcohol per capita has been increasing at an annual rate of 2.5 per cent (33) and in the United States at between 1.5 per cent and 2 per cent (34) A recent international study of mortality due to cirrhosis of the liver, a useful index of alcohol consumption, concluded 'The increase (between 1950 and 1971) was so great that it is possible to say that there is an epidemic of cirrhosis of the liver'.(35) The serving of alcohol at social functions is increasingly common and the number of people who drive to social functions is also increasing. Alcohol is implicated in 25 per cent of fatal road accidents in Britain, and in over 50 per cent in Canada and the United States. (36, 37) A 1974 study by the Canadian Ministry of Transport which administered breathalyser tests to a large random sample of Ontario drivers between the hours of 10 p.m. and 3 a.m. found that 27 per cent of the sample has been drinking and 6.4 per cent were over the legal limit.(38) On any given night in North America and Western Europe it is a fairly safe guess that millions of drivers who have been drinking are on the road, and the number is increasing. A tiny fraction of this number actually gets caught by the police. In North America, estimates of the chances of being caught while over the limit range from 1 in 2500 to 1 in 7500, and in France one estimate puts the risk of conviction for impaired driving at 1 for every 175 000 kilometre driven.(39) Obviously, if there were the political will, laws could be passed and enforced that could virtually eradicate the offence of drunken driving. But the will appears to be lacking. In Britain the introduction of the breathalyser illustrated the sort of impact that could be made on the accident figures. But as Figure 7.8 shows, the effect began to wear off almost immediately.

The lack of political will to crack down hard on drinking and driving can be explained in part by the fact that a very large number of normal people would get caught in the crack down. But another part of the explanation is that, statistically speaking, driving while moderately impaired is not dangerous. Every day many millions do it and only a few thousands come to serious grief. Although a blood alcohol concentration of 80 mg per 100 ml has been estimated to double the risk of an accident, for any particular trip home after a party this represents a doubling of an extremely small probability. In 1973 the fatal accident rate in the

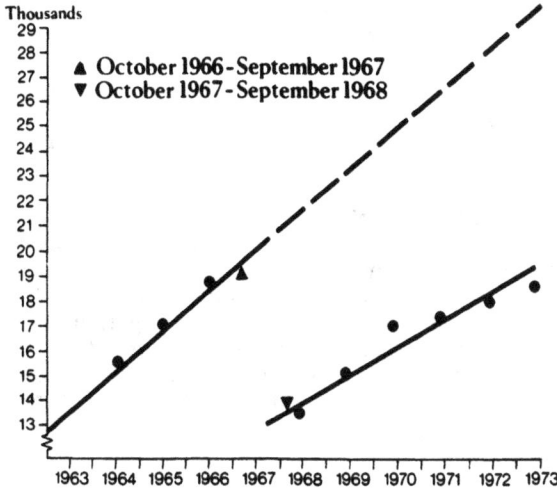

Figure 7.8 Trends in numbers of car involvements in fatal and
serious accidents between 10 p.m. and 4 a.m.
Source: 'Road Accidents Great Britain 1973', HMSO.

United States was 4.2 per 100 million vehicle miles. If that risk
were doubled the chances of surviving a 1-mile trip home would
still be pretty good; on average one could expect to make the trip
safely about 12 million times for every fatal accident and even if
one were driving with double the legal blood alcohol concentration
the odds would still be over 1 000 000 to 1 against a fatal accident.
(40) Many experienced drinking drivers are good judges of their
driving abilities while moderately impaired, and have 'clean'
licences to prove it.

The great difficulty in reducing the number of road deaths in
a country with a very high death toll such as the United States
is, paradoxically, that driving is so extremely safe. It is very
difficult to persuade someone that he ought to forgo something he
wants, such as a few drinks, or driving a few miles an hour over
the speed limit, for the benefit of reducing his risk of a fatal accident by one 12 millionth for every mile driven. And it is difficult
to get people to take seriously the idea that doing something that
increases the risk of a fatal accident by this imperceptible magnitude is a 'crime'. Certainly it is unlikely that the existing legislation dealing with impaired driving would have received the
political support that it did had it been promoted on the basis of
risk reductions of this magnitude.

In Sweden and Finland where drinking and driving is treated as
a serious crime punishable by a jail sentence, close to a half of all
prison sentences are for drinking and driving offence.(41) In the
Netherlands, where it is also commonly punished by incarceration,

there is a special prison for these 'criminals' to protect them from contamination by 'real' criminals.(42) It has been observed that wherever the penalties for impaired driving are severe it is common to find compensatingly low rates of enforcement and conviction.(43) There appears to be no statistical evidence to support the view that putting impaired drivers in jail is an effective way to reduce the road death rate. Certainly the positions of Finland, Sweden and the Netherlands on Figure 7.2 are consistent with this lack of evidence.

It is the conjuction of extremely small risks per vehicle kilometre with extremely large numbers of vehicle kilometres that produces the total death toll which society judges unacceptable. And yet the only way of making substantial inroads on this total is by legislation that requires drivers to behave in a way that they judge individually to be excessively prudent. Figure 7.9 shows that as car ownership increases it becomes more difficult to make substantial reductions in the number of deaths per motor vehicle. Current safety measures appear to be straining the democratic limits of driver perfectibility.

Another part of the explanation of the apparent willingness of society to tolerate its road death toll lies in the biased sample of society that is consulted. It has already been noted that in Western Europe and North America most legislators and a large and influential proportion of the electorate are drivers. Drivers have an obvious tendency to view the accident problem from the driver's perspective. Although some drivers are killed through no fault of their own, it is, nevertheless, drivers who cause most traffic accidents. Drivers, because of the possibilities they have for taking evasive action when threatened by other drivers, and because of the freedom they have to choose the risks that they take, have much more control over their own safety than either passengers, pedestrians or cyclists. Thus drivers have much more reason to be content with the level of risk entailed in driving. In Britain in 1976 only 25 per cent of traffic fatalities were drivers of cars, buses or lorries. The remaining 75 per cent were pedestrians, cyclists, motor-cyclists or passengers. For every car driver killed in a fatal accident involving a car, there were more than three non-drivers killed; for every lorry driver killed in a fatal accident involving a lorry there were more than twelve other people killed.(44)

It can be argued, with some justification, that the law ought not to be employed to protect people from themselves. This libertarian philosophy applied to road accidents unfortunately deals with only about one-quarter of the problem. But those responsible for judging what society will and will not tolerate display a preference for modifying the behaviour not of the drivers but of their potential victims. In a discussion of the numbers of children killed on the roads in Britain the government, in an official discussion paper, declares 'The seeds of these accidents are sown in traditions of independence and freedom; sometimes also in thoughtlessness and lack of care. Parents and children alike need to be educated in the

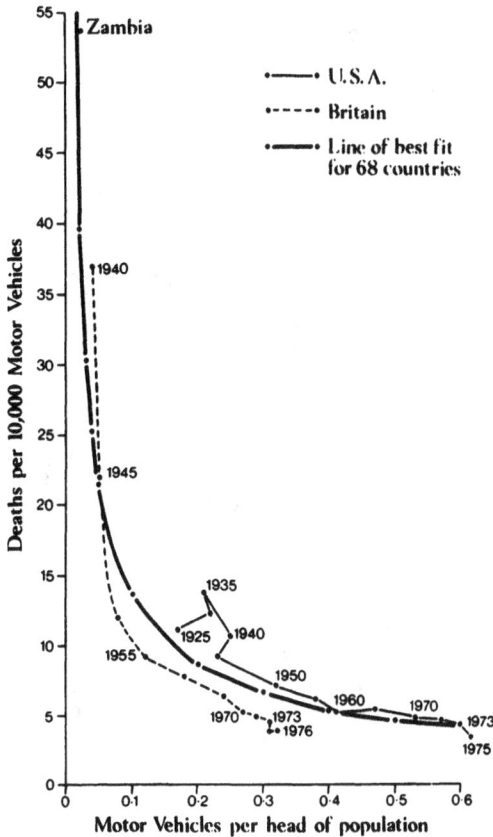

Figure 7.9
Sources: 'Road Accidents Great Britain 1973', 'Transport Statistics
Great Britain 1966-76'; 'Statistical Abstract of the US
1977'; and Smeed, R. J. and Jeffcoate, G. O. (1970),
Effects of Changes in Motorization in Various Countries
on the Number of Road Fatalities, 'TEC', vol. 12, no. 3,
pp. 150-1.

dangers and the means to reduce them.'(45) In Detroit, also, the
response to traffic accidents involving children has been to curb
their traditional freedoms. The result has been described by
Bunge:
Where could the Indian children travel across Fitzgerald's
landscape [a neighbourhood in Detroit]? Everywhere. By
the time of late farm days the fences were spreading yet the
children could still safely use most of the roads and wander
in considerable open spaces like Holman's Woods. Today the

children can move almost nowhere. They are more and more
caged. Expressway fences and property fences continue to
go up. These fences are often built with the excuse of protecting the children from the machines, especially the automobile, but it is the machines which are being given the
space taken from the children.(46)

## SUMMARY

Fatal accidents are the top prizes in the automotive lottery. With
every lapse of concentration and error of judgment a driver pur-
chases another ticket. The chance of any one ticket being drawn
during any 1 mile of driving in the United States is about 1 in 25
million. The tickets are very easy to come by. They can be
bought by driving after a few extra drinks, by running on a
worn tyre, by going a bit too fast, by driving when sleepy, or
angry, or preoccupied, by showing off, by skimping on maintenance etc. etc., but the chances of any particular ticket being a
winning one are negligible. Just as in a conventional lottery, the
number of tickets drawn is a small but fairly constant proportion
of the number of tickets bought. In 1973 in Britain on an average
day twenty winning numbers were drawn, in Canada eighteen, and
in the United States 152.

There is a widespread agreement that it would be desirable to
reduce the number of prizes. But so far no effective way has been
found, and there is disagreement about what ought to be tried
next. Some advocate social engineering methods for altering the
risks that people will tolerate in order to reduce the number of
tickets that are bought. Others prefer physical engineering methods for increasing road and vehicle safety, thereby reducing the
proportion of tickets that are drawn. Each school of thought has
some modest achievements to its credit but each also has compelling reasons for doubting the effectiveness of the solutions preferred by the opposing school. If the probability of a ticket being
drawn is reduced then, the risk-compensation theorists argue,
drivers will respond by buying more tickets. The physical engineering school counters by noting that people resent and resist
interference with their freedom to buy the tickets, especially in
the extremely small denominations in which they are sold.

The random number generator that produces each day's winning
numbers has millions of moving parts, lethal machines erratically
piloted by fallible humans. Each day these machines zoom, criss-
cross, and occasionally collide with each other, with pedestrians,
or with roadside obstacles. Individually these collisions are unpre-
dictable, but the number of fatal winning tickets drawn each day
behaves with remarkable consistency. Both the physical and social
engineers can claim some credit for reducing the number of collisions in the generator, per moving part, but both have failed to
halt the increase in the total number of tickets drawn.

Every new car driver automatically joins the lottery and becomes a purchaser of tickets. The growth in the number of drivers in the twentieth century has made the automotive lottery the biggest game of chance in the Western world.

REFERENCES AND NOTES

1  Kahn, H. (1976), 'The Next 200 Years', William Morrow & Co., p. 168.
2  'Hansard', 6 April 1977.
3  Carruthers, M. (1974), 'The Western Way of Death', Pantheon Books.
4  'World Health Statistics Report', vol. 27, no. 8, 1974, The Ten Leading Causes of Death for Selected Countries in North America, Europe and Oceania, 1969, 1970 and 1971.
5  Peltzman, S. (1975), The Effects of Automobile Safety Regulations, 'Journal of Political Economy', vol. 83, no. 4.
6  Havard, J. D. (1975), The Drinking Driver and the Law: Legal Countermeasures in the Prevention of Alcohol-related Traffic Accidents, in R. J. Gibbins et al., 'Research Advances in Alcohol and Drug Problems', vol. 2.
7  'Encyclopedia Americana'.
8  'Transport Statistics Great Britain 1964-1974'.
9  'Statistical History of the United States from Colonial Times to 1970', Basic Books, 1976.
10  'World Almanac 1978'.
11  Campbell, B. J. (1973), Accident Proneness and Driver License Programs, Paper presented to the First International Conference on Driver Behaviour Organized by the International Drivers' Behaviour Research Association, Zurich, Switzerland.
12  Smeed, R. J. (1968), Variations in the Pattern of Accident Rates in Different Countries and Their Causes, 'Traffic Engineering and Control', vol. 10, no. 7, pp. 364-71, and, with Jeffcoate, G. O. (1970), Effects of Changes in Motorization in Various Countries on the Number of Road Fatalities, 'TEC', vol. 12, no. 3, pp. 150-1..
13  Eames, W. G., Lee, S. N. and Fell, J. C. (1970), State of the Art - Motor Vehicle Accident Studies, 'International Automobile Safety Conference Compendium', Society of Automotive Engineers Inc.
14  In Britain before 1954 a person who was injured in a road accident and subsequently died was recorded as 'killed' if he died within two months; in 1954 this time period was changed to one month.
15  Department of the Environment (1976), 'Transport Policy: A Consultation Document', HMSO, vol. I, Table 5.
16  'Statistical Abstract of the US 1977', Table 1050.
17  'Hansard', 4 July 1973, col. 556.
18  Plowden, W. (1973), 'The Motor Car and Politics in Great Britain', Penguin.
19  Hamer, M. (1974), 'Wheels Within Wheels', Friends of the Earth.

20  Davies, R. O. (1975), 'The Age of Ashphalt', J. B. Lippin-cott, NY.
21  Schneider, K. R. (1972), 'Autokind versus Mankind', Schoc-ken Books, NY.
22  Cameron, J. (1974), How the Interstate Changed the Face of the Nation, in S. J. Hille and R. F. Poist (eds), 'Transpor-tation Principles and Perspectives', Interstate Printers and Publishers, Danville, Illinois.
23  Davies (1975), op. cit.
24  US Department of Transportation (1975), '1974 National Trans-portation Report, Current Performance and Future Prospects', p. 292.
25  Wilde, G. J. S. (1976), 'The Risk Compensation Theory of Accident Causation and its Practical Consequences for Acci-dent Prevention', paper presented at the Annual Meeting of the Österreichische Gesellschaft für Unfallchirurgie, Salzburg.
26  Borkenstein, R. F. (1975), Problèmes d'application, juge-ments et sanctions en ce qui a trait aux loi sur l'alcool et la sécurité routière, 'Toxicomanies', vol. 8, no. 1, Table 2.
27  Peltzman (1975), op. cit.
28  Wilde (1976), op. cit.
29  Eames et al. (1970), op. cit., p. 1259.
30  'Drinking-Driving in the Province of Ontario', a report to the Provincial Secretary for Justice by the Inter-Ministerial Com-mittee on Drinking and Driving, August 1974, p. 4.
31  Eames et al. (1970), op. cit., p. 1258.
32  Ferrence, R. G., and Whitehead, P. (1977), Impaired Driving and Public Policy: an Evaluation of Proposed Countermeasures, 'Blutalkohol', vol. 14, p. 3.
33  'Drinking-Driving in the Province of Ontario', op. cit., cal-culated from Table A3.
34  Peltzman (1975), op. cit.
35  Masse, L. et al. (1976), Trends in Mortality from Cirrhosis of the Liver 1950-71, 'World Health Statistics Report', vol. 29, no. 1.
36  Borkenstein (1975), op. cit.
37  Eames et al. (1970), op. cit.
38  'Drinking-Driving in the Province of Ontario', op. cit., p. 5.
39  Wilde, G., personal communication.
40  Probabilities based on Figure 1 of 'Drinking-Driving in the Province of Ontario'.
41  Ferrence and Whitehead (1977), op. cit.
42  Dijksterhuis, F. P. H. (1975), The Specific Preventative Effect of Penal Measures on Subjects Convicted for Drunken Driving, 'Blutalkohol', vol. 12, pp. 181-91.
43  Ross, H. L. (1975), L'efficacité des lois sur l'ivresse au volent en Suede et en Grande-Bretagne, 'Toxicomanies', vol. 8, pp. 54-76.
44  'Road Accidents Great Britain 1976', HMSO, Tables 4, 11 and 12.
45  Department of the Environment (1976), op. cit. p. 76.

46 Bunge, W. (1971), 'Fitzgerald: Geography of a Revolution', Schenkman Publishing Co., Cambridge, Mass., p. 242.

# Part II

# PRACTICE:
the British way

# 8 THE POLICY ENVIRONMENT

By the mid-1970s transport planning in Britain had run into serious trouble. Airport planning had reached an impasse. The airlines, the airport operators and the planners were adamant that a major new airport was needed in the south-east of England; and everywhere that they contemplated building it they met fierce and implacable opposition from local residents. But perhaps the most serious trouble encountered by transport planners, certainly the most widely publicized, was the disruption that took place at motorway inquiries. The government responded by announcing two inquiries into inquiries. Both the form and content of the motorway planning process were reviewed. The review of form was conducted by the Department of Transport and the Department of the Environment in consultation with the Council on Tribunals. It resulted in the 'Report on the Review of Highway Inquiry Procedures'.(1) The review of the content was conducted by what became known as the Leitch Committee and resulted in the 'Report of the Advisory Committee on Trunk Road Assessment'.(2)

Figure 8.1 describes the formal steps that are taken before the opening of a new road in Britain. Although the procedures followed in the planning of roads in other countries and of other major civil engineering projects in Britain are different in detail they are similar in principle; they all involve a planning sequence in which a highly generalized concept of need is progressively refined into a highly specific construction project.

The Leitch Committee's job was to inquire into steps 1, 5 and 9, the steps concerned with forecasting and assessment; the procedural review was concerned primarily with steps 13 and 16, the public inquiry stages.

All the little boxes in Figure 8.1 sit within a larger box containing policy. Policy is an ill-defined mixture of statements in government White Papers, ministerial pronouncements, and tacit assumption. It forms the atmosphere within which the procedures of assessment and inquiry are conducted. It cannot, by well established convention, be questioned by these procedures. Nor was it questioned by the inquiries into what was wrong with these procedures.

So long as policy was generally accepted as reflecting a consensus about the objectives of transport planning, then the procedures outlined in Figure 8.1 worked fairly well. All those involved in the planning procedures breathed the same air of policy, unselfconsciously, unthinkingly and unquestioningly. Up until the early 1970s there was, arguably, a strong consensus in

129

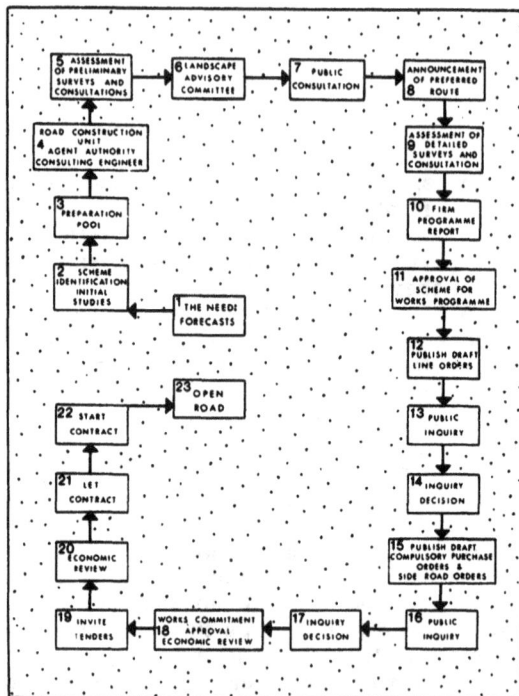

Figure 8.1 Stages in the evolution of a trunk road system
Source:   Based upon Table 1.3 of the 'Report of the Advisory
          Committee on Trunk Road Assessment'.

Britain, especially among transport planners about the desirability
and inevitability of the mobility transition. A congestion-free,
high-mobility future for all was the goal they were all working
towards.

This consensus has now broken down, and a new one has yet to
emerge. The government discussion paper(3) that preceded the
government's 1977 transport policy White Paper (4) candidly admitted that the government did not have a coherent transport
policy. The White Paper argued that public transport should be
improved, that energy should be conserved, and that there should
be an absolute reduction in the nation's dependence on transport.
At the same time it vigorously promoted the form of transport, the
car, that most effectively undermines public transport, that consumes energy extravagantly, and that, by its encouragement of
land-use sprawl, increases the nation's dependence on transport.
The Conservative government that has succeeded the authors of
the White Paper adheres equally strongly to the same set of contradictory objectives.

What both reviews conspicuously failed to recognize was that the troubles at inquiries were a reflection not primarily of technical or procedural failures, but of a breakdown in the established consensus. The policy atmosphere within which transport planning is now conducted has become opaque with ambivalence and acrid with conflict.

REFERENCES AND NOTES

1 'Report on the Review of Highway Inquiry Procedures' (1978), HMSO, Comd 7133.
2 Department of Transport (1977), 'Report of the Advisory Committee on Trunk Road Assessment', HMSO
3 Department of the Environment (1976), 'Transport Policy: A Consultation Document', 2 vols, HMSO.
4 Department of Transport (1977), 'Transport Policy', HMSO, Cmnd 6836.

# 9 FORECASTING I:
## prophecy

> The prophets they prophesy falsely, and the priests bear
> rule by their means; and my people love to have it so: and
> what will ye do in the end thereof?
>
> Jeremiah, v. 31

In Jeremiah's day forecasting was called prophecy. The prophets'
authority derived from a claimed special relationship with the div-
inity. Hence their occupation has also been known as divining.
Today their authority derives from science. What will be is no
longer the will of the Almighty but the working out of impersonal,
atheistic laws of nature.

But little else has changed. For the layman both the will of God
and the laws of nature are shrouded in mystery. It is as difficult
for him to dispute an authority deriving from a claimed special
understanding of the latter as from a claimed special relationship
with the former - until the day of reckoning. If the world does
not come to an end on the day prophesied, the prophet's authority vanishes. Hence successful practising forecasters from
before the time of the Delphic Oracle to the present day have cul-
tivated the fine art of ambiguity.

It is no coincidence that one of the most popular methods employed by the social sciences today for forecasting socio-economic
change is called the Delphi Method. This is a method much like
that employed by insurance companies for spreading large risks.
Large numbers of experts pool their forecasts in such a way as to
make it impossible to attach much of the blame for getting it wrong
to any one of them. But this protection is needed only rarely
because the forecasts are usually so well protected by a cocoon of
ambiguous conditional clauses that it is difficult to prove any
liability at all.

At the moment, however, there are some rather large claims
pending. The decision by those who bear rule to commit vast
sums of the people's money to building the world's fastest passenger airplane was based on prophecies (of how much it would
cost to develop, build and fly Concorde, of the amount of noise it
would make, and of the numbers who would fly in it) that have
proven resoundingly false. In 1971 Sir George Edwards, Chairman of the British Aircraft Corporation, undismayed by the
awkward fact that he could find no one who wished to buy Concorde, was arguing that 'the replacement of the ageing subsonic
jets used on ranges above 2000 miles, plus an annual growth rate
of 5 per cent could well require 1500 Concordes and Concorde
development aircraft to be in service by the end of the century'.(1)

Substantial sums of money have also been committed to building the longest bridge in the world, across the Humber, on the basis of forecasts of traffic that now appear unlikely ever to materialize. The Central Electricity Generating Board has expensive generating capacity substantially in excess of its needs as a consequence of placing its faith in forecasts that proved false. Britain's airport planners narrowly averted similar embarrassment. In 1970 the Roskill Commission predicted that by 1991 the volume of traffic through London's airports would be 137.9 million passengers, and on the basis of this forecast argued that the need for a new airport was urgent. The most recent official forecasts for 1990 give a range from 65.9 million to 89.4 million and these also will almost certainly be proven gross overestimates (see Chapter 10).

TRAFFIC FORECASTING

The forecasts that at present serve to justify the largest expenditure of public money are the road traffic forecasts. Figure 9.1 describes the recent track record of the most important component

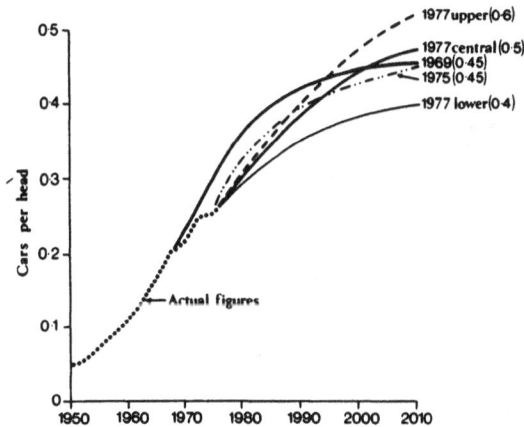

Figure 9.1 Car ownership forecasts
Sources: for 1969: Tulpule, A. H. (1969), 'Forecasts of Vehicles and Traffic in Great Britain', Transport and Road Research Laboratory, LR288; for 1975: Tanner, J. C. (1975), 'Forecasts of Vehicles and Traffic in Great Britain: 1974 Revision', Transport and Road Research Laboratory, LR650; for 1977: Tanner, J. C. (1977), 'Car Ownership Trends and Forecasts', Transport and Road Research Laboratory, LR799. Figures in brackets after the year of forecast indicate the ultimate level of car ownership assumed.

of these forecasts, the car ownership forecasts. Their justifying function can be seen at its most direct in decisions to build new roads. The government's policy, enunciated both at public inquiries into road schemes and in Parliament (11 June 1975) is that 'roads are provided when it is believed the traffic that will be generated over the next fifteen years will necessitate roads of a particular standard'. The forecasts are set alongside the Department of Transport highway design standards in order to determine what size of road will be needed to serve the traffic expected in the 'design year', i.e. fifteen years after the road has been opened. They are also used as input to the Department's cost-benefit analysis programme; whatever benefit a new road is deemed to have for an individual motorist is multiplied by the number of motorists expected to use the road. Thus a high traffic forecast will justify building more and bigger roads than a low traffic forecast.

The traffic forecasts also serve as the basis of the transport planning component of structure plans, and are used as input to a variety of other transport planning models, such as the new noise and visual intrusion models. They served as the principal justification for the scale of expenditure proposed in the 1970 White Paper 'Roads for the Future'. In presenting the most recent version of his forecasting model to the Royal Statistical Society in 1977, Mr John Tanner, for many years the government's principal traffic forecaster, stated 'If we did not believe that there was something inevitable about traffic growth, then we would not be producing the sort of forecasts that we have been.'(2) Traffic growth is viewed as an autonomous process that will continue into the indefinite future. Within constraints set by the ability of the public to provide the funds, the road provision and expenditure targets of the long-term national transport programme are strongly influenced by the forecasts of the rate of traffic growth. The traffic forecasts have become an established part of the climate of opinion within which transport problems are evaluated.

Their influence extends beyond the realm of transport planning. The 'energy crisis' that has provoked all the current concern with energy forecasting is primarily an 'oil crisis'. Oil is the country's most important source of energy and its supply has recently come to appear extremely precarious. As noted in Chapter 5, the single most important use for oil, and the use for which it is most difficult to find substitutes, is road transport. The Department of Energy's forecast of oil consumption by road transport is produced by making assumptions about future volumes of road traffic and about the fuel efficiency of the vehicles comprising this traffic. The assumptions about future volumes of road traffic used are the traffic forecasts of the Department of Transport. The Department of Energy, like the Department of Transport, treats its energy consumption forecasts as estimates of a demand that must be met by the provision of more capacity.(3)

Regional planners also see themselves as the willing servants of the inevitable. The following passages, taken from the 'North East

Lancashire Development Plan: Report on Sub-regional Development', are characteristic of planning in the early 1970s:

> Para. 7.22  It is pointless to refute the argument that the
> building of more roads will further encourage car usage and
> result in additional losses of passengers to train and bus
> services. The net effect will be to encourage the contraction
> of public transport services and to require large increases
> in the subsidy required for their operation because of falling
> revenue.

> Para. 11.14 ... in Blackburn for instance it is envisaged
> that current road programmes and redevelopment schemes
> might mean the disappearance of 860 shops – about 21% of
> all the town's retail floorspace and 38% of all shopping units.

> Para. 6.58  To some extent we regard the process of concen-
> tration as inevitable. It would be vain to think that, by
> attempting to obstruct investment in larger, more centralized
> facilities, local planning authorities could thereby ensure an
> even spread throughout the existing structure: in most
> fields modern, high-grade facilities are simply not viable at
> the local level. Larger and more specialized facilities will be
> an indirect result of the authorities' own action in improving
> transportation.

It is conceded by the planners that their activities influence
events and yet at the same time 'to some extent' they consider the
process of which they are a part to be inevitable. Tanner con-
fronts this paradox in his reply to critics of his Royal Statistical
Society paper 'who accuse me of making implicit assumptions
about future policies and thus usurping the role of the Government'.

> My view on this is that provided that we continue to live in
> a democracy, over the extended time-scale with which we
> are concerned the transport situation will tend to reflect
> underlying economic, technological and social forces, and
> that its broad outline is therefore a subject for prediction
> as much as for planning. I do not believe that extreme
> policies could or would be sustained over extended periods
> of time, unless for example the energy situation demanded
> it.(4)

Mr Tanner and the development planners manage to coexist comfortably with this paradox because they are persuaded that the
process which they are at the same time forecasting and abetting
is a desirable one. In North East Lancashire, for example, it is
transforming facilities considered obsolete and low-grade into ones
that are 'modern' and 'high-grade'. In Tanner's view traffic
growth is what the people have chosen and, so long as democracy
survives, will continue to choose. In his earlier work he refers to
the 'saturation level of car ownership', the statistical culmination
of the growth process, as a 'target level' and states that 'the
current target level represents the current desire of people in an
area for car ownership'.(5)

There is no doubt evidence that many people who do not own cars would like to have them, especially those dependent on deteriorating public transport. But no evidence has been produced by the Department of Transport, or its forecasters, that people either individually or collectively desire to live in a country saturated with cars. The 'saturation level' of car ownership, suggesting as it does a situation in which the country cannot absorb any more cars, is a particularly apt term to describe the ultimate state to which the Department of Transport's planners appear to aspire. The Department has never attempted to describe what it would be like to live in such a society, and then to ask people if this is what they want. It has merely assumed that it would be desirable, and asserted that it is what people want.

### THE FORECASTING METHODS

Road traffic is dominated by cars and this dominance is expected to continue. Thus the most important part of the road traffic forecast is the forecast of car traffic. This forecast is in turn based on forecasts of car ownership per capita, the distance travelled per car, and population. Of these it is car ownership that accounts for by far the largest part of the forecast increase in traffic. In recent years there have been a number of attempts by forecasters to summarize the way in which the various forces, economic, social and technological, combine to determine the rate at which car ownership will increase.

The earliest encapsulation of these forces has its roots in Tanner's forecasting work in the early 1960s. It looks like this

$$y = \frac{s}{1 + \dfrac{(s - y_0)\, e^{-rs(t - t_0)/s - y_0}}{y_0}}$$

This was further refined by Mr Tanner in 1974 (6) to look like this

$$y = \frac{s}{1 + \dfrac{s - y_0}{y_0}\left(\dfrac{i}{i_0}\right)^{-bs}\left(\dfrac{p}{p_0}\right)^{-cs} e^{-as(t - t_0)}}$$

And in 1977 (7) to look like this

$$y = \frac{s}{1 + [a(t - t_0) + b \log i + c \log p]^{-n}}$$

Finally, in 1978 (8) a rival group of forecasters produced a model that looked like this

$$P_{1+} = \frac{s}{1 + e^{-a}(i/P)^{-b}}$$

| | | | |
|---|---|---|---|
| $y_0$ | = car ownership in base year | $i$ | = income per head |
| $t$ | = year for which car ownership is forecast | $p$ | = cost of motoring |
| $t_0$ | = base year | $P$ | = purchase price of cars |
| $n$ | = years after base year | | |

All of this, to the uninitiated, looks very impressive, almost
impressive enough to be what it purports to be - the distilled
statistical essence of the combined economic, social and technological forces that determine the rate at which car ownership will
grow in the future. But it is really extraordinarily simple. On the
left of the equal sign we have either y or p1+; these are what the
forecasters are attempting to forecast. y is the number of cars
per head of population. p1+ represents the proportion of households owning one or more cars. On the right of the equal sign we
find an expression with a complicated denominator and an extremely simple numerator. The numerator, s, is the saturation
level of car ownership. Below the line we find that all the models
have something in common; they all begin with 1+. To the right
of the + sign things become a bit complex. But given time and/or
an increasing gross domestic product this complexity disappears.
All the expressions to the right of the + sign get smaller and
smaller as time passes or GDP increases until ultimately they become small enough to ignore completely, in which case we are left
with

$$y \text{ or } p_{1+} = \frac{s}{1+0} = \frac{s}{1} \text{ or, simply, } s$$

All of these equations when represented by a graph produce S-
shaped curves and all arrive at the same point, the saturation
level, in the end. So long as the forecasters can agree on the
saturation level, their various models will produce roughly similar forecasts. In Figure 9.1 the 1968 forecast is based on model
(i), the 1975 forecast on model (ii) and the 1977 forecast on
model (iii). With each revision the forecast has been corrected to
allow for the fact that the previous forecast produced an overestimate. For 1969 and 1975 a saturation level of .45 was used, in
1977 three different levels were used, .4, .5 and .6, and it can
be seen that they result in substantially different forecasts. The
solid curved line of Figure 9.2 shows the most recent model,
model (iv). (More will be said below about the broken line.) It
produces an estimate of cars per household rather than cars per
person but aside from making comparisons with the previous
models rather difficult this is not a very significant difference.
Also, instead of showing the way car ownership is expected to
vary with time it shows the way in which it is expected to vary
with income. But since income is assumed by the forecasters to
grow steadily over time this is not a significant difference either.
In the long term if the same value for the saturation level is fed
into models (iii) and (iv) then they will produce exactly the same
forecast. In its most recent memorandum containing the official
traffic forecasts used for purposes of road planning ('The National Traffic Forecasts: Interim Memorandum 1978') the Department of Transport presented a compromise between the forecasts
produced by these two models. Towards the end of the forecast
period the two models give remarkably similar estimates.
    The saturation level, s, is clearly the key number. It has an

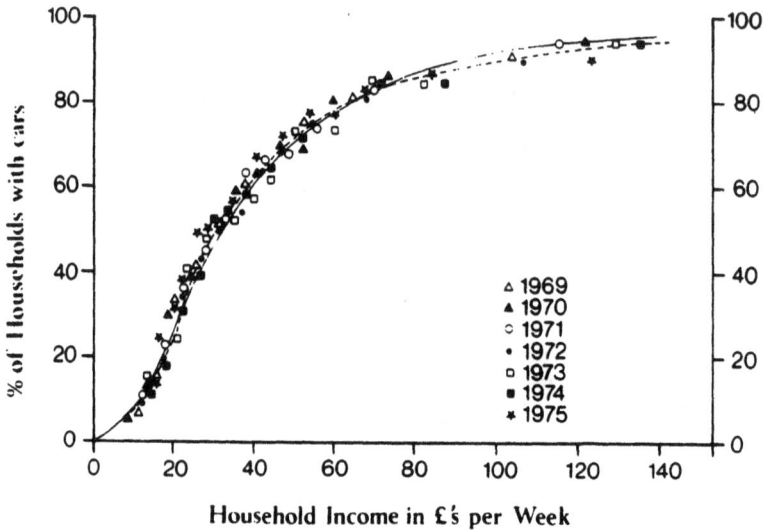

Figure 9.2 The relationship between car ownership and household income

Source;  Bates, J. (1978), 'A Disaggregate Model of Household Car Ownership', Departments of the Environment and Transport, Research Report 20.

interesting history. Committed to the idea that it was a fit subject for scientific prediction the forecasters attacked their data with the conviction that, if traffic growth were a lawful process, in the scientific sense, then a study of the previous history of this growth would yield the secret of its future course. Figure 9.3 illustrates the relationship that, for a number of years, was assumed to reveal this secret. If the growth process followed an S-shaped logistic curve, and if all parts of the country could be assumed to be following exactly the same growth process, then all counties in Britain ought to lie on the straight line of Figure 9.3, and ought, over time, to slide slowly down the line until they reached the bottom, at which point growth everywhere would have ceased. If these hypotheses were right, fitting a line through the dots on the graph and extending it to the horizontal axis ought to reveal the level of car ownership at which growth would stop; the point where this line crosses the horizontal axis is called the zero growth intercept. For a number of years this method constituted the principal means by which the saturation level was estimated. An elementary test of the method shows that over time it produced hopelessly inconsistent results. The estimate of the saturation level it produced varied enormously depending on the period of years for which the car ownership levels and

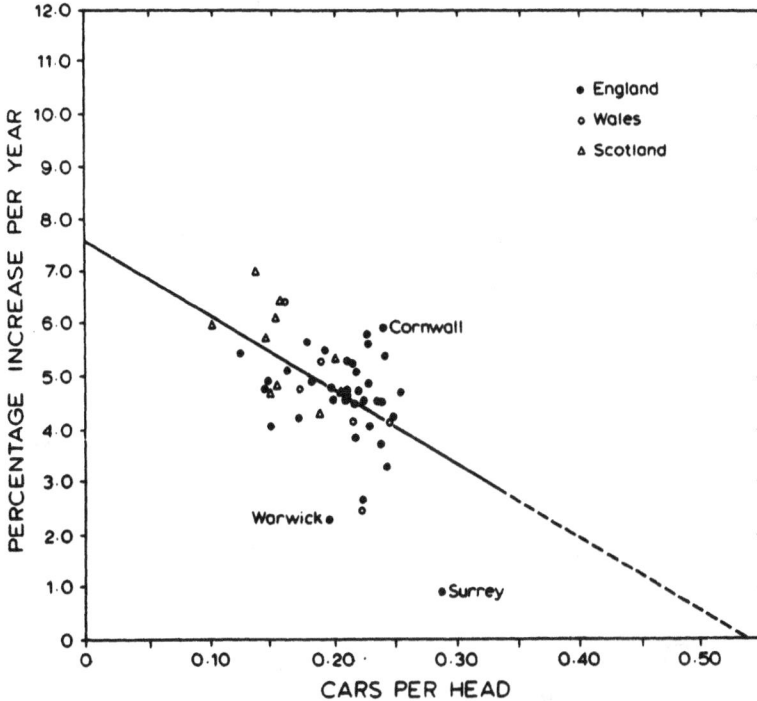

Figure 9.3 Method for estimating the ultimate 'saturation' level of car ownership
Source: Taken from TRRL Report LR 543.

and growth rates were calculated. Figure 9.4 displays some of this variety.

At one stage the logistic curve on which this method was based acquired for the Departments of Transport and the Environment almost magical properties. It became a universal law of human and biological behaviour. In defence of their method they argued at the public inquiry into the M16 Motorway (now M25), 'If growth in car ownership behaves like other social and biological phenomena, it will lie on a logistic growth curve ...' This was pure superstition. The car ownership growth process has not followed a logistic curve in the past and there is no reason why it should in the future either. When pressed, the Departmental spokesman could not produce a single example of a social or biological phenomenon that followed this curve with anything remotely approaching sufficient precision to make it a useful basis for forecasting.

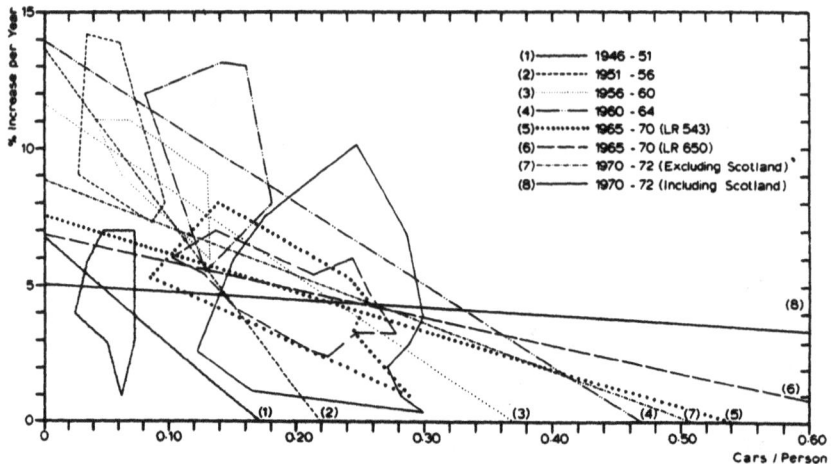

Figure 9.4 The inconsistent results of the method illustrated by
Figure 9.3.
(The irregular polygons form the outer boundary of the data
points for the relevant time period, e.g. polygon 5 and the dotted
line represent the data and the line of best fit of Figure 9.3)


## PROPHETIC DISCOMFITURE

The growing opposition to motorways in the 1970s was accompanied
by growing criticism of the traffic forecasting and evaluation
methods used to justify them. As a consequence, an advisory committee, known after its chairman as the Leitch Committee, was
appointed by the government to examine the criticisms and recom-
mend improvements. The Committee's first recommendation was a
piece of public relations advice worthy of the Delphic Oracle itself:
'We conclude our review of the Department's procedures by em-
phasizing that uncertainty in forecasting is inevitable and it is
important that such uncertainties should be exposed to public
scrutiny ... the Department should never put itself in the position
of appearing to defend a single figure as if it were uniquely correct.'(9)
  What will be will be, uniquely. What is uncertain is not what
will be, but our knowledge of it. A decision to build a road of a
specific size, in a specific place, at a specific time, amounts to a
commitment to accommodate a specific amount of traffic. A decision
to build, amounts to a commitment, albeit in the face of uncertainty, to a figure within a narrow range set by the highway
design standards. The key word in the Leitch Committee's oracular
pronouncement is 'appearing'. The Committee does not say who
should be responsible for defending the figure upon which the
road building commitment rests, only that it should not appear to
be the Department of Transport.

Two additional recommendations of the Leitch Committee are also worthy of note: 'The Department should as soon as is practicable move away from the extrapolatory form of model currently used towards basing its forecasts on causal models,' and, 'We believe, that since Government has undertaken to consult the public on the decisions it takes about trunk roads, it has a fundamental obligation to ensure that the methods of assessment it adopts are fully comprehensible.'

The Committee objected to the Department of Transport's long established method of forecasting for the sorts of reasons discussed above:

In any extrapolatory model all that is known is the behaviour of the relevant variables over a past time period. Future data are of course unknown. The question then arises how best to fit a curve to the known data and what trajectory it should follow into the future - in this case up to the assumed saturation level. The logistic curve is only one of the many forms of curve that it is possible to fit to the known data. Thus there is nothing uniquely right about the S-shaped curve that determines the level of car ownership in future years.

The first traffic forecasts to emerge from the Department of Transport after publication of the Leitch Committee Report are contained in Research Report 20 'A Disaggregate Model of Household Car Ownership'.(10) The fact that it was not produced by the Transport and Road Research Laboratory, the traditional source of official forecasts, is an indication that the Leitch Committee's criticism of the previous forecasts is being taken seriously by the Department. A second indication is found in the introduction of 'Report 20' where the new model is described as a departure from the previous method of extrapolating past trends. The new approach, it says, 'seeks to identify the variables which influence car ownership'.

It also contains what appears to be a response to the Leitch Committee's strictures on comprehensibility; it contains an appendix 'to provide an illustration for those unfamiliar with the techniques'. The appendix is five pages long and begins quite encouragingly:

A brief and highly simplified outline of the likelihood analysis approach follows below. It is perhaps best illustrated by a simple example. Suppose a modeller has a number of observations of some random variable, X say, and wishes to describe the probability density of X in terms of some function $f(x,\Theta)$, which includes some unknown parameter $\Theta$. The modeller supplies the form of the function $f$, and now has to choose a value of $\Theta$ on the basis of the set of observations, say $(x_1 \ldots x_n)$. The 'likelihood function' of the set of observations is defined as

$$L(\Theta) = \prod_{i=1}^{n} (fx_1,\Theta)$$

and so on for five pages. That, if anyone asks, is how they drew
the solid curved line displayed in Figure 9.2.

Here is the conclusion of the Report:

provided income is adjusted for changes in the general level
of prices and additionally for changes in the relative price
of cars, FES data [Family Expenditure Survey data] suggests that the relationship between car ownership and income
has remained constant to an approximation sufficient to the
requirements of a medium term forecasting model. Hence it
is suggested that changes in the level of car ownership may
be predicted by predicting real income growth and change
in the level of car prices (subject of course to the usual
caveats inherent in assuming any relationship to remain
fixed in the future).

Caveat emptor indeed. What does it all mean? Simply that they
have found a fairly strong relationship over the period 1969 to
1975 between a household's income and the likelihood that it will
own a car. This in itself is not a novel discovery. It has long
been fairly widely known that rich people tend to own more cars
than poor people. The advance claimed for the 'likelihood function'
is that it describes this relationship more accurately than ever
before. Figure 9.2 and equation (iv) above represent 'Report
20's' principal finding. As can be seen, the relationship is indeed
a strong one and the solid curved line drawn through the symbols
on the graph appears a fairly reasonable generalization of the
relationship. All the business about likelihood functions, to which
the Report is largely devoted, is concerned with an extremely involved way of calculating the path of this line of 'best fit'. A
naive inspection of Figure 9.2 suggests however that a better line
could be fitted by drawing it free hand. It can be seen that in the
top half of the income range there are no points above the solid
line. In the middle of the range there are too many points above
the line. At the bottom there are, as at the top, too few. The
curved broken line represents a freehand attempt at improving
the accuracy of the line of best fit.

This line fits the data better because, unlike the solid line, it
is not based on a function supplied by a modeller that represents
a mathematical preconception about the nature of the relationship.
The pen that drew the freehand line was free to follow where the
data led. The difference between the two lines is of course sufficiently small to be dismissed as a mere quibble. But it is to such
quibbling that 'Report 20' is largely devoted. There are pages
and pages of befuddling graphs, tables and equations devoted to
justifying the use of the new 'computationally more complicated'
maximum likelihood method in place of the old highly complicated
linear regression method, on the basis of differences in the line
of best fit no greater than that displayed in Figure 9.2.

The use of the new method for making forecasts rests on two
improbable assumptions: that a relationship that existed for the
period 1969-75 between car ownership, household incomes and car
prices will persist to the year 2005, and that it is possible to

predict with reasonable accuracy what will happen to incomes and prices, and hence to car ownership levels, between now and 2005. If Figure 9.2 could be expected to retain its shape unaltered for thirty years, and if all households could be expected to ascend the income ladder in an orderly fashion, and if the rate of this ascent could be estimated with reasonable accuracy, then it would be possible to estimate future levels of car ownership with confidence.

The most remarkable finding of 'Report 20' is that 'it is only the purchase price of cars which affects the level of car ownership, and not the cost of running the car'. This, the authors concede, may surprise some people or to put it in their scientific language, '... may seem to some people counter-intuitive'. Figure 9.5 displays some of the information upon which the Department of Transport and the Transport and Road Research Laboratory relied for developing models (iii) and (iv). Both models have a cost variable in their denominators. p in model (iii) refers to 'motoring costs' that include the costs of both buying and running a car. P in model (iv) refers only to the cost of buying cars.

In estimating the rate at which car ownership will increase in the future the authors of 'Report 20' disregard completely any anticipated rise in the price of petrol. The reason they give is that the car running cost index, which is dominated by the price of petrol, 'performed badly' in their statistical analysis. Intuition

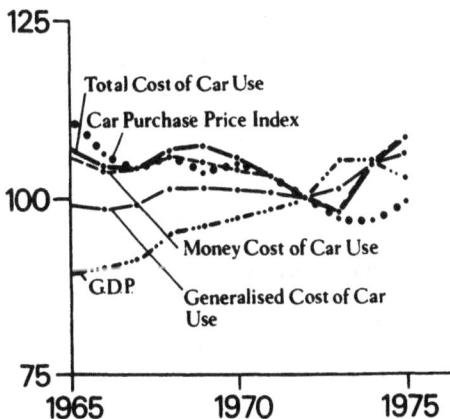

Figure 9.5 Total cost of car use and car purchase price indices. (The 'generalized cost of car use' includes an estimate of the value of time spent travelling)

Sources: Tanner, J. C. (1977, 'Car Ownership Trends and Forecasts', Transport and Road Research Laboratory, LT799; and Bates, J. (1978), ' A Disaggregate Model of Household Car Ownership', Departments of the Environment and Transport, Research Report 20.

suggests a number of explanations for this bad behaviour. Compared to the magnitude of the change anticipated in the future, the car running cost index did not change very much during the period covered by Figure 9.2; it averaged 1.05, and ranged between .99 and 1.13. One would not expect to find a sudden change in car ownership levels as a result of modest and somewhat erratic changes in the price of petrol over such a short period of time. Also, in a period of rapid inflation establishing the 'real' price of anything is rather difficult. The discrepancies between the different series of cost indices shown in Figure 9.5 give an indication of the problem of finding an accurate measure of costs. After 1973 when the price of petrol increased sharply, although by very little in comparison with the currently projected increases, the increase in car ownership virtually stopped. But since incomes declined at the same time it was difficult statistically to separate the effects of these changes and hence easy to dismiss the effect of the rise in petrol prices.

When it comes to estimating the effect that future petrol price increases might have on car use, as distinct from car ownership, the Department of Transport assumes it to be trivial. In 1976 the average car travelled 14 290 kilometres. The most recent forecasts allow for a range about this, in the year 2000, from 13 470 to 14 730. The memorandum accompanying the most recent forecasts is misleading about the way in which allowance has been made for future increases in the price of petrol. In discussing its car ownership forecasts it says 'the assumptions about fuel price and GDP now used are less favourable to traffic growth'. But the new car ownership forecasting model is totally insensitive to changing fuel prices because it does not include them.


THE USUAL CAVEATS

Although 'Report 20' declares itself to be subject to the 'usual caveats' inherent in assuming any relationship to be fixed in the future, it says very little about them. It mentions two, more or less in passing, that are quite interesting.

The first is the assumption 'that cross-sectional elasticities apply over time'. This means that as all households climb the income ladder the proportion of their incomes spent on buying cars will alter in a way consistent with the curve of Figure 9.2. In defence of this assumption it is argued:

> Since high income households tend to put a high value on personal mobility, it seems intuitively likely that they will wish to own a car; even if the members are unable to drive themselves, they can afford to employ other people to drive them. A [saturation level] figure much lower than 0.95 would be harder to justify.

At present there are far more households in the bottom half of the income range in Figure 9.2 than in the top half. The employers of drivers tend to be found in the top half; the employed

drivers tend to come from the bottom half. Since all households
are assumed to grow richer over time the assumption that non-
drivers will acquire cars requires that no matter how wealthy
people become some of them will still be willing to work for others
as chauffeurs. As car ownership increases and public transport
declines the demand for chauffeurs can be expected to grow while
the supply can be expected to shrink. 'Report 20' does not give
any indication of the numbers of people who will require chauffeurs, but they are likely to be many. A report of a mass screening survey published in the British Medical Journal of 2000 people
aged between 40 and 65 found that one in ten had vision very
substantially worse than that required by the Department of
Transport in order to be eligible for a driving licence.(11) As
indicated in Chapter 6, the proportion of the population capable
of qualifying for a driving licence has an upper limit of around
60 per cent.

A second interesting assumption upon which the model rests is
that 'as households become richer they will use some of their
extra income for the purchase of more space'. This is a fairly
important assumption because in many parts of the country where
car ownership is low at present there is neither parking space nor
driving space for the increase in the car population envisaged by
the forecasts. Unfortunately, since all households are assumed to
grow richer over time this assumption requires either the creation
of more land or the taking of land from non-residential uses, most
notably from agricultural uses.

In 1975 there were approximately twenty-five cars for every
100 people in Britain. In 'Report 20' it is suggested that the saturation level will lie somewhere between fifty-nine and sixty-seven
per 100. The magnitude of the change in the nation's land-use
pattern that would be required to accommodate such an increase
is a question that deserves more attention than it receives in
'Report 20'. In a 1974 forecasting report from the Transport and
Road Research Laboratory it was acknowledged that the change
required to accommodate even forty-five cars per 100 people would
be considerable: 'Such a level would of course require considerable
changes in the nature of our society including the geographical
distribution of the population...'(12)

In answer to those who expressed doubts about the *desirability*
of these required changes, the Department answered they were
*possible*: 'In Connecticut, which has a higher population density
than this country, the ownership rate is presently 0.55'.(13) What
is being compared to Britain by the Department is an American
state having approximately one-twentieth the population of Britain
and one-twentieth the area. Although it has a similar population
density it has a very different population distribution, its largest
city, Hartford, has a population of 162 000, approximately the size
of Brighton. The state is largely a suburban part of Megalopolis,
the giant urban/suburban sprawl that stretches from Washington to
Boston. In the book entitled 'Megalopolis'(14), Jean Gottmann describes its problems of transport as follows:

146 *Practice: the British way*

the growth of Megalopolis owes much to the automobile but
highway traffic jams are beginning to strangle city activities
and to take the pleasure out of driving a car. At the same
time cars contribute to the ruination of other means of
transportation, made more necessary than ever by the massive tidal currents of people and goods.
Peter Hall, in his book 'The World Cities', suggests that it is not
an example that knowledgeable planners would seek to emulate.
Here is evolving a type of urban area without parallel in
eastern North America: an importation from the universal
sprawl of Los Angeles. It depends almost wholly on the automobile, for a finely developed railroad net, or even adequate
express bus transportation, is no longer economic. The
commuter bound for Manhattan must drive long distances to
a suburban railhead; his wife needs a second car for the
long journey to the shopping centre.... Exurbia tends to be
occupied only by the most fortunate members of American
society... but rich as these people are, the communities are
in a chronic state of anxiety... here, in pure anxiety, lies
the origin of exurbia. It is not something that anyone appears
to want; it is produced by fear.
David R. Meyer in 'Urban Change in Central Connecticut' re-
inforces this impression.
The dispersal of population at low densities, of manufacturing at scattered sites, and of planned shopping centres
means that mass transit is rapidly declining as a feasible
component of the transportation system of central Connecticut.... The prospect is that by 1980 local bus service
will be almost nonexistent.... Because of the dispersal of
population and employment, auto driver trips are increasing
at a faster rate than population growth as more and more
activities can be reached only by car.... The end result
for the poor is to be almost totally immobilized in the industrial cities. In contrast the nonpoor will be served by a fine
network of highways and express commuter mass transit
service that will make it possible to live in pastoral environments at ever further distances from white collar employment
centres in Hartford and New Haven.
Both Hartford and New Haven, Connecticut's two largest cities,
participated in the immensely destructive wave of urban rioting
that swept the United States in the summer of 1967.
Mr Tanner argues that only undesirably 'extreme' policies can
prevent Britain from ultimately achieving the level of car ownership that Connecticut has now. Connecticut provides a useful
guide to the 'considerable changes in the nature of society' that
he has acknowledged would be required to achieve the growth that
he forecasts: a continued sapping of what life remains in the
nation's villages, the closing down of most of what remains of the
public transport system, and further decanting on a massive scale
of both jobs and people from inner cities into suburbs.

## PRIESTS AND PROPHETS

The authority of those who bear rule in a democratic country ought to derive from the support of those who elect them to office. But the future to which we are being committed by present government policies has never been put to the democratic test. There has never been an attempt to test the strength of informed support for a society saturated with cars. The authority for government actions committing us irrevocably to the consequences of growth derives largely, as it did in Jeremiah's time, from the forecasters. The forecasts are simply covert government policies for which neither the forecasters nor the government will accept responsibility. Growth in perpetuity is prophesied, those who bear rule strive mightily to make the prophecy come true, and a great many people in complete ignorance of the consequences doubtless love to have it so.

'And what will ye do in the end thereof?' This is a difficult question to raise, let alone to answer. The Department of Transport's prophecies cannot be discussed at the public inquiries where they are presented as justification for its road building proposals because of their national import. In the 'Report on the Review of Highway Inquiry Procedures' the Department explains 'local inquiries are unsuitable for examining technical issues, such as methods of trunk road assessment which have a national impact'. Equally, they cannot be discussed in Parliament because they are too technical. This was demonstrated very convincingly by Lord Avebury in a debate in the House of Lords. In an attempt to illuminate the inadequacies of the traffic forecasts he read out the mathematical formula at the heart of the forecasting model. The reply by the Department's spokesman was 'I do not profess to understand or to be able to translate the mathematical formula referred to by the noble Lord, Lord Avebury. I am grateful for his explanation and I hope that, when I read it tomorrow in "Hansard" it will be clearer to me what they are getting at'.(15)

The 'Report on the Review of Highway Inquiry Procedures' (16) acknowledges that there is a problem: 'technical matters must not be immune from rigorous examination by an independent body'. This is the answer it provides to the problem: 'The new Standing Advisory Committee, to be chaired by Sir George Leitch, which is being set up by the Secretary of State for Transport, will have a continuing responsibility to monitor developments in methods of technical assessment'.

The 'Report of the Review' would have us believe that considerable progress has already been made: 'As a result of the Leitch Report, revised national traffic forecasts have been issued by the Department of Transport on which it will be the Department's policy to rely.'

Reliance on its traffic forecasts is virtually the only policy that the Department of Transport has. Converting the temporary advisory committee into a 'standing' one does nothing to ensure that the forecasts will be subjected to rigorous independent scrutiny.

It will in fact make such scrutiny much more difficult. The Secretary of State is merely appointing additional prophets to advise him. The committee members are his appointees and thus clearly not independent. The newly appointed standing advisory committee may well dispute with the Department's established prophets, but after it has stood awhile, and advised awhile, it will become totally incapable of subjecting forecast that incorporate its own advice to independent scrutiny.

One suspects that the priests of Jeremiah's day were no more willing to subject the means by which they bore rule to independent scrutiny. In the conduct of human affairs there is much that has not changed in 2500 years.

## REFERENCES AND NOTES

1  Edwards, G. (1971), 'Flight International', 23 December 1971, p. 997.
2  Tanner, J. C. (1978), Long-term Forecasting of Vehicle Ownership and Road Traffic, 'Journal of the Royal Statistical Society', 141, Part 1, pp. 14-63.
3  (a) Department of Energy (1978), 'Energy Policy: a Consultative Document', HMSO, and (b) 'Energy Forecasting Methodology', Energy Paper No. 29, HMSO.
4  Tanner (1978), op. cit.
5  Tanner, J. C. (1962), Forecasts of Future Numbers of Vehicles in Great Britain, 'Roads and Road Construction', 40.
6  Tanner, J. C. (1974), 'Forecasts of Vehicles and Traffic in Great Britain: 1974 Revision', Transport and Road Research Laboratory, LR650.
7  Tanner, J. C. (1977), 'Car Ownership Trends and Forecasts', Transport and Road Research Laboratory, LR799.
8  Bates, J. (1978), 'A Disaggregate Model of Household Car Ownership', Departments of the Environment and Transport, Research Report 20.
9  Department of Transport (1977), 'Report of the Advisory Committee on Trunk Road Assessment', HMSO.
10  Bates (1978), op. cit.
11  Stone, D. H. and Shannon, D. J. (1978), Screening for Impaired Visual Acuity in Middle Age in General Practice, 'British Medical Journal', 2, pp. 859-63.
12  Tanner (1974), op. cit.
13  Department of the Environment (1976), The National Traffic Forecasts, vol. II, Paper 7, in 'Transport Policy: A Consultation Document', HMSO.
14  Gottmann, J. (1961), 'Megalopolis', MIT Press, p. 12.
15  'Hansard' (Lords), 25 February 1976, col. 804.
16  Departments of Transport and the Environment (1978), 'Report on the Review of Highway Inquiry Procedures', HMSO, Cmnd 7133.

# 10  FORECASTING II: policy

Forecasts, the forecasters insist, are required for planning. The forecasts of the Departments of Environment, Transport, Energy and Trade are growth forecasts and the planning purpose they serve is growth. In 'Energy Policy' it is spelt out clearly; the objective of energy policy is to ensure 'adequate and secure energy supplies'. 'Adequate' the paper explains, 'implies that they should not be a constraint on economic growth'. The Department of Transport's version of this policy is somewhat stronger. The 'first objective' of transport policy is not just the removal of constraints on growth but the positive promotion of growth: 'to contribute to economic growth and higher national prosperity'. The Department of Trade's version of the same policy is presented in its White Paper 'Airports Policy': 'air transport has been a high growth industry' which 'provides a substantial benefit to the economy' and therefore 'ways should be sought for meeting efficiently the growing consumer demand for air transport'.(1)

All growth processes encroach on their environments. If the environment is finite, the growth of one thing implies the destruction or the pushing out of the way of something else. According to orthodox economic theory, as resources grow scarcer their price ought to rise higher, thereby inhibiting demand. With economic growth comes a growing demand for land for 'development' - for roads, factories, suburbs, shopping centres, airports, and power-stations; in the early 1970s agricultural land was lost to such purposes at a rate of more than 100 square miles (25 900 hectares) a year. Other costs more difficult to measure are also incurred. Havens from the ubiquitous noise of internal combustion and jet engines become harder to find, treasured landscapes are disfigured, communities are disrupted. But since none of these costs, economic or non-economic, enters the forecasters' equations, the environment of the growth processes they are forecasting is assumed to be free.

Although there exists no offical statements adopting this assumption as government policy, nevertheless it is treated by the forecasters as government policy. Tanner puts it this way: 'Clearly government actions have some influence on future levels (and could if necessary determine them), and therefore any forecasts must be based on explicit or implicit assumptions about the policies of future governments.'(2) They must also be based on assumptions about the policies of present governments.

The necessity of policy assumptions presents the forecaster with a serious problem. In its transport policy consultation document

149

published in 1976 the government confessed that 'by common con-
sent, we still lack a coherent national transport policy'. It still
lacks one. If a forecaster searches in 'Transport Policy' (the
White Paper published a year later)(3) for assumptions upon which
to base car ownership forecasts, he will find that the government
aims to improve public transport, conserve energy, and reduce
the nation's dependence on transport. These aims, it might be
argued, imply a reduction in car ownership levels. But he will
also find that the government aims to promote both economic
growth and the car industry - aims that imply an increase in car
ownership levels.

If a forecaster's employer refuses to provide him with a coherent
set of policy assumptions, the forecaster can either refuse to do
the job, or he can provide his own assumptions. The road traffic
forecasters have chosen to provide their own. Perhaps the most
noteworthy of the assumptions provided by Tanner is that 'traffic
will not be deliberately and substantially restrained for environmental or social reasons'. The forecasts of traffic growth that are
derived from this assumption are used at public inquiries to justify building more roads, and it is argued by the proponents of the
roads being inquired into that not to build them would restrain
the growth of traffic. But the extra future traffic that would be
restrained is entirely hypothetical. It is traffic that would materialize only if future travellers were not confronted with the social
and environmental costs of their travel. Future traffic predicted
by a model that ignores these costs could be described as traffic
generated by means of a concealed social and environmental subsidy, a subsidy that grows larger as traffic increases.

In presenting his forecasts to a meeting of the Royal Statistical
Society, Tanner observed that criticism of his forecasts had 'often,
not surprisingly, ... come from sources which dislike the policy
implications of the forecasts'. His apparent implication was that
such criticism ought therefore to be discounted because it was
prejudiced. He was taken to task by Professor Stone in the discussion that followed: 'I think that Mr. Tanner should pause
to examine the other side of that coin which is that the acceptance
of his forecasts may come from organizations that *like* their policy
implications'.(4)

Some of the organizations that like their policy implications are
fairly obvious: those who make cars, those who build roads, those
who sell petrol, and so on. But the most important of the organizations that can be presumed to like both the forecasts and their
policy implications is the Department of Transport. It is primarily
a department of road transport and contains within its ranks large
numbers of people whose working lives have been devoted to
planning the expansion of the country's road network. Despite
the fact that civil servants are supposed to be the servants of
their political masters it would clearly be unreasonable to expect
of them an overnight conversion to policies that called a halt to
this expansion or, even worse, that implied that it had been mis-
guided.

Road building projects have long gestation periods during which those who work on the projects acquire an interest in seeing them through. In 1978, for example, the Department of Transport estimated that more than £815 000 had been spent in preparing its proposals for widening a contested 1-mile stretch of the Archway Road in north London.(5) This sum included £380 000 for property acquisition, £250 000 for consultants' fees and £135 000 for conducting public inquiries. The figure did not include departmental staff costs. Including staff costs and converting all the sums - some of which were spent thirty years ago - to 1978 prices would produce an expenditure estimate well in excess of £2 million - before construction had even begun. This is an exceptional case; it nevertheless provides a measure of the interest that those who have worked on the project might be expected to have in seeing it through to completion.

One further obvious and influential group whose members like the policy implications of the government's growth forecasts is the group made up of the forecasters themselves. Everyone enjoys being proved right and the forecasters can only be proved right if they have guessed right about the future policies upon which they have built their models. Since most members of this group believe growth to be desirable, their growth forecasts become a mathematized form of wishful thinking. A good example of this phenomenon is found in the work of a prominent forecasting institution called the Henley Centre for Forecasting, which regularly produces forecasts of the rate at which the economy will grow. Inside the front cover of each of its publications is the following statement of the Centre's raison d'etre: 'Through its contributions to better decision-taking, a scientific approach to forecasting can play its part towards the attainment of faster economic growth'.(6)

## FORECASTS OF GOODS TRAFFIC BY ROAD

The Leitch Committee concluded that 'more attention must be paid to the forecasting of commercial vehicle traffic'. The Committee's terms of reference specifically enjoined it 'to review the Department's method of traffic forecasting, its application of the forecasts and to comment on the sensitivity of the forecasts to policy changes'.(7) But despite its terms of reference and despite the fact that evidence highly critical of the commercial vehicle forecasts was submitted to it, the Committee paid the matter no attention whatsoever.

One possible explanation of this omission is that the commercial vehicle forecasts (8, 9) were simply too embarrassing to bear thinking about. Because the forecasts assumed first that the number of tonne-kilometres of freight moving about the country would, for the indefinite future, be firmly tied to gross domestic product, and secondly, that gross domestic product would grow at 3 per cent 'in perpetuity', it was rather easy to demonstrate their absurdity.(10)

The Department of Transport now has a new method for forecasting road freight traffic. The details of the new method, at the
time of writing, are secret. The Interim Memorandum, the current
source of 'official' forecasts for planning purposes, says 'The
freight forecasts included in this memorandum are based on recent
and still continuing work within the Department, the results of
which will be published shortly.'(11) These new forecasts, it
explains, 'are based on forecasts of the output of major industrial
sectors, of the proportion of that output moved by road, and of
the average length of haul'. The new methods predict a faster
increase in the number of tonne-kilometres of freight. But they
also assume a faster increase in the size of lorries. These two differences, the Memorandum observes, largely cancel each other so
that the traffic forecasts, i.e. the estimates of vehicles-kilometres
produced by the new method, are very similar to the old. Whereas
the old method rested on a forecast of an aggregate measure of
economic activity, GDP, the new method requires separate forecasts for all the major components of the economy. The only thing
likely to be gained by this complicating of the forecasting method
is an increase in jobs for forecasters and a decrease in the number of people who will understand the result. The old method,
statistically speaking, was really quite serviceable. As Figure
10.1 shows, the correlation between tonne-kilometres of freight
moved in Britain and GDP has been very strong. An examination
of the ratio of tonne-kilometres to GDP for thirteen other countries over a number of years suggests that, while there is considerable variation among countries, within each country, over
time, the ratio is very stable.(12) If this ratio is known, and if it
remains stable, and if one can predict a country's GDP for a given
year, one can predict reasonably accurately the quantity of
freight traffic for that year.
    It is sometimes suggested that the ratio of tonne-kilometres to
GDP will not remain stable but can be expected to decrease over
time. Because a tonne of computers is worth more than a tonne
of bricks, and because high value low weight commodities such as
computers are forming a growing share of GDP, it is argued that
GDP could continue to grow while the quantity of freight moved
increased by very little, if at all. If this were to happen the
straight line shown in Figure 10.1 would start to bend over and
level off. In theory it could even turn down while GDP continued
to increase. In practice, however, there is no sign of this beginning to happen, even in countries considerably wealthier than
Britain. For example, Figure 10.2 shows that in the United States
where per capita incomes are almost twice those in Britain the
relationship has remained strong and linear for over twenty years.
It appears that any increase in the value per tonne of production
has been offset by an increase in the distance each tonne travels.
This is because production of the highest value goods is very
highly specialized, draws upon dispersed and remote sources of
raw materials, involves many intermediate steps between raw
materials and final consumers, and is sold to widely dispersed
final consumers.

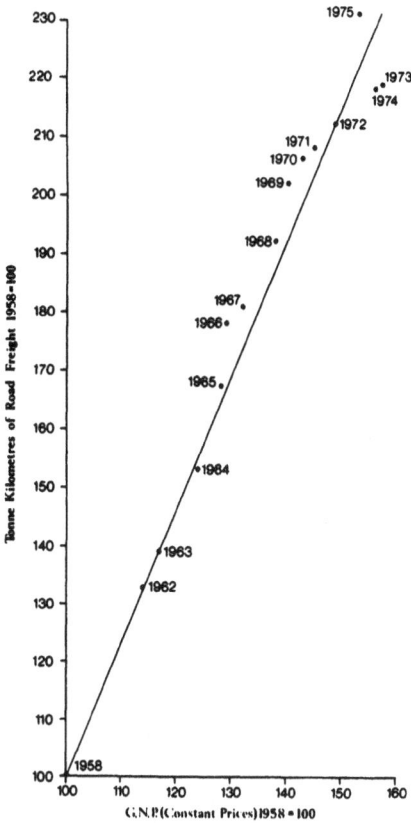

Figure 10.1 Great Britain

The previous lorry forecasts, like the present ones, rested on
assumptions about the rate at which lorries would increase in size.
The assumptions previously used for the 'optimistic' high growth
forecasts implied that towards the end of the forecast period
'the carrying capacity of a typical 2010 lorry would be toward the
upper end of the range of today's lorries'.(13) It is an awkward
mathematical fact that as the average lorry size approaches the
maximum lorry size, the minimum will approach the average, and
all lorries will tend to the same size. The most recent forecasts
assume that there will be no increase in the maximum weights of
lorries. While all the new assumptions about vehicle fleet composition and average carrying capacity are still secret, the Interim
Memorandum does reveal that the proportion of the vehicle fleet
assumed to consist of vehicles having more than two axles is
greater than assumed before. Hence, by the year 2010 the typical
vehicle is assumed to be even closer to the size of the largest
juggernaut currently permitted on the roads.

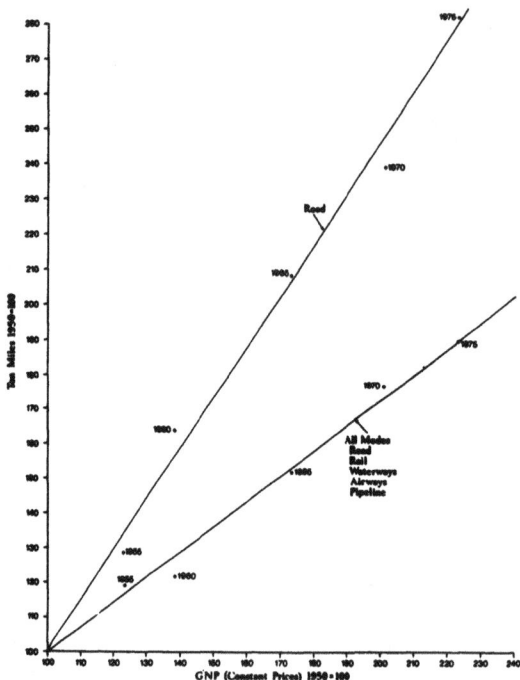

Figure 10.2 Growth of intercity freight traffic 1950-75 USA

Heated controversy exists about the possibility of the railways relieving the roads of a part of their burden of freight. Doubtless a considerable amount of the long-distance freight that now travels by road could, at little or no extra cost, be diverted to rail. But the heat of the road versus rail debate obscures a much more important aspect of the freight problem. In 1976 British Rail carried 20.4 thousand million tonne-kilometres of freight traffic. The previous 'optimistic' high growth road freight forecast estimated that traffic would increase from 87 000 million tonne-kilometres in 1972 to 313 000 million by 2010. Thus the increase in road freight traffic forecast was more than ten times the total traffic carried by rail. The new method, we are told, gives higher estimates, although we are not hold how much higher. In the context of growth aspirations such as these the debate about diverting modest percentages of traffic from road to rail is at best an irrelevance and at worst a distraction from a much more urgent problem.

The new Department of Transport forecasting method, like the old, when projected far enough into the future presents the country with a stark choice between an absurd number of vehicles and vehicles of absurd size. There is an influential lobby within the

Department of Transport that favours a compromise between these choices. Clearly, the more quickly numbers grow the less quickly will their size become absurd, and vice versa. But this compromise involves raising the weight limit currently imposed on lorries. In October 1978 internal Department of Transport memoranda discussing how this might be done in the face of the present strong public opposition to larger lorries leaked out and were widely reported in the Press. It was clearly the view of the senior Department of Transport civil servants taking part in these discussions that opposition to larger lorries was 'irrational'. The civil servants concerned were pilloried in the Press for presuming to promote their own pro-road policies against the wishes of their political masters.(14) But a balanced growth of both lorry numbers and lorry sizes is arguably the least irrational way of coping with the freight consequences of their political masters' economic growth policies.

## AIR TRAFFIC FORECASTING

Air traffic is increasing more rapidly than traffic by any other mode. By 1976 it accounted for 14 per cent of energy consumed in the transport sector, compared with 78 per cent by road and 4 per cent each by rail and water traffic. Since 1950 the energy consumed by air travel has increased by 753 per cent, while that consumed by road traffic has increased by only 255 per cent, and that consumed by rail and water traffic decreased by 88 per cent and 4 per cent respectively.(15) During this period the number of journeys through British airports has increased by over 1600 per cent to more than 40 million a year, and the average journey length has also increased. For passenger travel particularly, air is the mode of transport with the greatest potential for further growth. As yet it is a mode of travel regularly used by only a small minority of the population; in 1975 in Britain one round trip by air was made for every six persons in the population. Air travel is also the mode of travel most dependent on oil; expenditure on fuel accounts for about 25 per cent of total airline operating costs,(16) and for the foreseeable future there are no alternative fuels to replace oil.

The following is a brief account of the model used for forecasting air traffic. It is taken without alteration from 'United Kingdom Air Traffic Forecasting: Research and Revised Forecasts' published by the Department of Trade in 1978.(17)

Estimated equations

1 UK Inclusive Tours (UKIT)
    a) UKIT = -5.54 + 1.0 UKPOP + 4.15 UKPCE - 2.44 UKITP
    b) UKIT = 15.85 + 1.0 UKPOP + 1.7 UKPCE - 4.63 UKITP
2 Other UK Leisure W Europe (UKSHL)
    UKSHL = -0.91 + 1.77 UKPCE

3 Other UK Leisure Rest of World (UKLHL)
  a) UKLHL = 4.51 + 1.5 UKPCE + 1.0 UKPOP - 2.37 UKLHLF
  b) UKLHL = 1.46 + 2.0 UKPCE + 1.0 UKPOP - 2.21 UKLHLF
4 Foreign Leisure W. Europe (FOSHL) (1962-73 equations)
  a) FOSHL = 4.52 + 2.0 WEPCE + 1.0 WEPOP - 2.6 FOSHLF
  b) FOSHL = -14.47 + 1.0 WEPOP + 3.5 WEPCE
5 Foreign Leisure Rest of World (FOLHL)
  FOLHL = 2.82 + 1.3 ROWPCE + 1.0 ROWPOP - 1.39 FOLHLF
6 UK Business - W. Europe (UKSHB)
  UKSHB = -2.61 + 1.97 WEGNP
7 UK Business - Rest of World (UKLHB)
  UKLHB = -2.34 + 2.42 ROWGNP - 0.92 UKLHBF
8 Foreign Business - w. Europe (FOSHB)
  FOSHB = -5.05 + 2.41 WEGNP
9 Foreign Business - Rest of World (FOLHB)
  FOLHB = -2.71 + 2.51 UKGNP - 0.70 FOLHBF

| | | |
|---|---|---|
| UKPOP | = | UK population |
| UKPCE | = | UK per capita consumer expenditure |
| UKITP | = | UK Inclusive tour price |
| UKLHLF | = | UK Long haul leisure fares |
| WEPOP | = | W Europe population |
| WEPCE | = | W Europe per capita consumers expenditure |
| FOSHLF | = | Foreign currency short haul leisure fares |
| ROWPOP | = | Rest of world population |
| ROWPCE | = | Rest of world per capita consumers expenditure |
| FOLHLF | = | Foreign currency long haul leisure fares |
| WEGNP | = | W Europe gross national product |
| ROWGNP | = | Rest of world gross national product |
| UKLHBF | = | UK Long haul business fares |
| UKGNP | = | UK gross national product |
| FOLHBF | = | Foreign currency long haul business fares |

The equations as quoted here refer to the logarithmic values of
the variables and the coefficients therefore have the properties
of elasticities.

Figure 10.3 The air traffic forecasting model of the Department
of Trade

Although this is, like the car ownership forecasting model, in-
timidatingly sophisticated to the layman, it is also in essence
extremely simple. With it, if you can predict accurately future
changes in the variables to the right of the equal signs, if you
can estimate accurately the parameters that relate these variables
to the different categories of traffic to the left of the equal signs,
and if the relationships defined by the parameters remain stable
or change in a predictable way, then you can predict the volumes
of traffic that will pass through the nation's airports in future
years.
  There are of course formidable difficulties associated with all

three ifs. The record of the forecasters in dealing with the most
basic of the variables (UKPOP) does not inspire confidence (see
Figure 10.5 on p. 161). The difficulties associated with estimating
the future gross national product of a place as large as the rest
of the world (ROWGNP) are likely to be at least as great. The
second if is undermined by what in the Department of Trade's
report is described as 'a major technical problem of ... multi
collinearity'. This is a statistician's way of saying that if traffic
has increased while at the same time incomes have increased, fares
have decreased and travel has become faster, it can be very difficult to identify accurately the causes of the traffic increase. The
third if, it is conceded in the report, rests on 'a considerable
element of judgment'.

Reduced to their barest essentials the Department of Trade's
judgment and hypothesizing involve feeding into its model the
assumptions that everywhere in the world people will grow richer
and the costs of flying will diminish.(18) The model responds to
this feeding by predicting that the flow of international traffic
through Britain's airports will increase from 30.2 million in 1975
to between 70.8 and 96.7 million in 1990.

Remarkable though the assumptions fed into the model are, even
more remarkable are the factors that are excluded from it. More
than a third of the traffic through London's airports belongs to
what is called 'the foreign leisure component'. This component is
predicted to grow from 10.2 million in 1975 to between 27.1 and
32.0 million in 1990. Commenting on the plausability of this forecast the report observes 'The most important need will be for
additional accommodation and the question arises as to whether it
is physically possible for this to be provided and, if it is, whether
the relative price of accommodation will rise.' This question has
not been incorporated into the forecasting model but the report
does speculate about how it might affect future traffic levels:

> The growth in air traffic forecast in this paper, even taking
> an optimistic view of the factors affecting the demand for
> total bedspaces in London, suggests that there will be
> increasing pressure, in the longer term, on hotel space in
> London. To generate the necessary additional investment it
> is likely that there would have to be quite considerable real
> increases in hotel tariffs, in order to provide a reasonable
> return on the capital employed. The real price increases
> necessary can be expected to have some impact on the gen-
> eral level of tourist demand. Little work has been done in
> this area and there are a number of factors which could mean
> that hotel prices, while affecting demand, do not greatly
> affect the number of passengers passing through airports.
> Nonetheless there is a clear possibility that this 'supply'
> effect could lead to some downward revision of the air traffic
> forecasts, particularly in the later years and at the high end
> of the range.

This is all that the forecasting report has to say about the probable impact on London of a threefold increase in foreign tourists

by 1990. It is all.that the report has to say about the impact of
the rising tide of tourism anywhere. There is no discussion of the
transformation in the character of London that would be required
to cope with the swollen foreign leisure component. There is no
intimation of the fact that many people in London already feel that
the pressure of visitors has changed the city's character for the
worse, or that existing levels of traffic are a cause of concern in
almost all the world's major tourist areas. There is no hint of the
fierce resentment of aircraft noise in the residential areas around
existing London airports, nor of the fact that after years of
searching, the airport planners have found nowhere to build the
new airport capacity that would be needed to handle the traffic
that has been forecast, nor of the fact that airport planners in
most of the world's major cities are confronted by similar problems.

In 1970 the Roskill Commission, which was inquiring into London's airport problems, described the way in which airport
planners approached their responsibilities: 'the rule-of-thumb now
current in the aviation world is that the average delay to passengers (caused by airport congestion) should only exceed 4 minutes
during the 30 busiest hours of the year'.(19) This rule of thumb
and a more elaborate version of it developed by the Roskill Commission are the airport planner's equivalent of the road planner's
design standards. They are applied to determine when, and how
much, additional airport capacity ought to be built to accommodate
the forecast growth in traffic.

If the demand for anything increases and the supply does not,
then the price rises to whatever level is necessary to bring demand
into balance with the supply. But Britain's airport planning is
based on forecasting methods that recognize no supply constraints
whatever, either within the airport perimeter, in terms of runways
or terminals, or beyond the perimeter in the areas for which the
travellers are destined. Demand is assumed to grow until limited
by global surfeit. The Department of Trade's report observes:

It is generally recognised that, particularly for leisure
travel, there are limits to the continued growth of air
travel even under favourable price and income conditions.
These will be imposed largely by limited time available for
travel and by constraints arising from demographic factors.
However, determining the point at which these limits start
to hold down traffic growth is by no means straightforward.

Because it is not straightforward the report says no more about it.

The Roskill Commission also worried briefly and inconclusively
about this problem. It suggested that, for no particular reason it
could think of, there might be an upper limit that would be reached
when every man, woman and baby made an average of six leisure
trips a year and every business person made an additional six
business trips a year. However, the Commission could envisage no
upper limit to the number of foreigners who might visit the country. Figure 10.4 shows the Roskill Commission forecasts and the
most recent Department of Trade forecasts. The most recent forecasts are much lower and venture to predict much less far into the

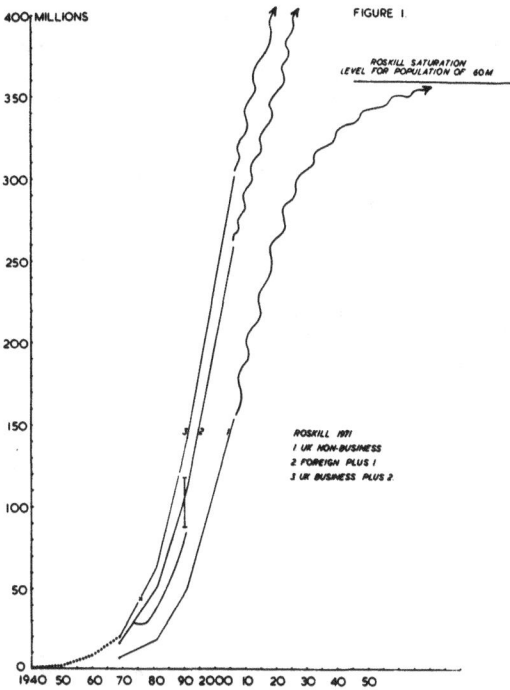

400 MILLIONS

FIGURE 1.

ROSKILL SATURATION
LEVEL FOR POPULATION OF 60 M

350

300

250

200

150

ROSKILL 1971
1 UK NON-BUSINESS
2 FOREIGN PLUS 1
3 UK BUSINESS PLUS 2

100

50

0
1940 50 60 70 80 90 2000 10 20 30 40 50

Figure 10.4
Line 3 represents the total traffic forecast by the Roskill
Commission for Heathrow, Gatwick, Luton and a third, unspeci-
fied, London airport. London area airports plus Manchester Airport account for approximately 75 per cent of the total traffic
through British airports. x indicates the total traffic for 1976.
The vertical bar at 1990 represents the range in total traffic forecast in the 1978 White Paper.
    The elongated J running from 1973 to 1990 represents the
British Airport Authority's response to the 1973 energy crisis. It
is a forecast of traffic for London area airports made in 1975. It
illustrates the Authority's conviction that 'normal growth' would
be delayed for three years.

future, but they rest on the same assumptions of unconstrained
growth. Such growth is viewed by the forecasters as the norm,
and any departure from it is considered a temporary aberration;
in a 1975 Civil Aviation Authority Report (20) commenting on the
abrupt halt to growth that occurred after the 1973 oil crisis it was
asserted that 'the effects of the present fuel and economic difficulties will be to delay normal passenger growth for a period of
three years'.
    Projecting the 'normal' post war growth trends in tourism to the

end of this century Herman Kahn concludes that 'it seems reasonable to assume that by the end of the century tourism will be one
of the largest industries in the world, if not the largest'.(21) The
British government is determined to safeguard Britain's share of
this industry. In 1969 the government established the British
Tourist Authority and national tourist boards for England, Scot-
land and Wales, all under the authority of the Department of Trade.
They were charged with 'responsibility for encouraging people to
visit and take holidays in their respective areas, and for encouraging the provision and improvement of tourist amenities and
facilities'.

Joining the current debate about whether London's share of the
industry is already too large, Michael Meacher, Parliamentary
Under-Secretary in the Department of Trade at the time, rebuked
those who publicly voice their resentment of the tourist influx for
failing to appreciate our economic dependence on tourism.(22) He
reminded those who spend their holidays in Torremolinos that it
is the British tourist industry that helps to pay their way. Doubtless those in Torremolinos who complain about the influx of English
tourists are reminded that it is tourism that helps to pay for their
holidays in England. Taking in one another's washing, Mark Twain
observed, is a precarious way of making a living. Taking in one another's tourists in a world of impending energy shortage seems
equally so.

Even if energy supplies were assured, however, there would
still come a time when the idea of encouraging foreigners to overrun Britain's 'attractions' in order to earn the money to permit
Britons to overrun their 'attractions' would lose its appeal for
almost all foreigners and Britons alike. But the appeal of the idea
is likely to persist longest for those whose livelihoods or political
fortunes depend on the tourist industry.


TIME, GDP, ENERGY AND TRAFFIC

Traffic growth, according to the models used to forecast it, is
inexorable. As noted in Chapter 9, the various models of the
Transport and Road Research Laboratory all assume that over
time car ownership will approach saturation level. The newest
model (model (iv) in Chapter 9) treats income rather than time as
the key variable. But this change is more apparent than real because income is assumed to increase relentlessly with time. Just
how relentlessly is emphasized by the statement accompanying the
forecasts used by the Department of Transport up until 1978 'The
central forecasts derive from an underlying GDP growth from 1973
onwards of 3 per cent per annum in perpetuity'.(23) The words
'in perpetuity' have been dropped from the statement accompanying
the most recent forecasts, perhaps as a concession to those who
argued that it unnecessary to forecast quite so far into the future,
but the Department's faith in the inexorability of economic growth
remains. The central GDP growth assumption, i.e. the assumption

about the rate at which I (income) in equation (iv) will increase,
has been modified to 3.5 per cent per annum up to 1985, and then
to 2.5 per cent per annum 'thereafter'. In the 1977 transport
White Paper, 'in perpetuity' is replaced by 'as far ahead as we can
see'.(24)

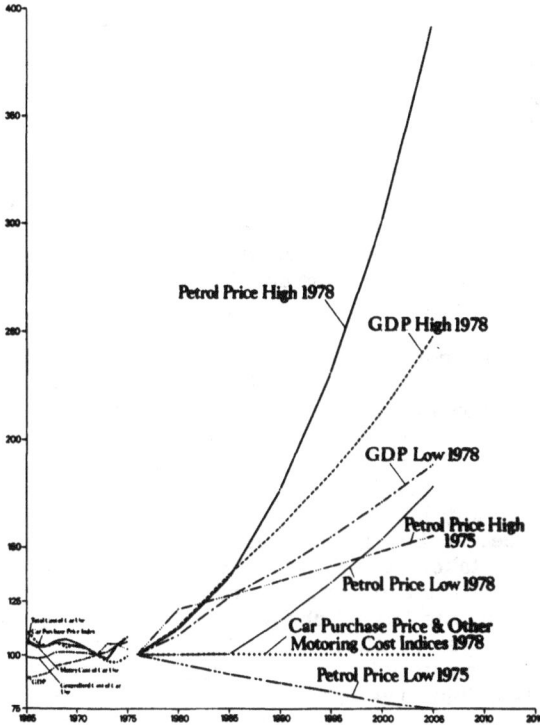

Figure 10.5 Assumptions about future trends of motoring cost
indices and GDP used for making the most recent forecasts.
(Figure 9.5 is represented in shrunken form in the lower left-hand
corner of the graph.)

Figure 10.5 places the Department of Transport's assumptions
about the growth of GDP as well as about the future behaviour of
the other critical indices alongside the behaviour of these indices
for the period of 1965-75. As has been noted above, the Department of Transport's traffic forecasts are used by the Department
of Energy as one of the most important components of its forecasting model. Levels of traffic, energy consumption and economic activity are all very closely related. Hence in the Department
of Energy's 'Energy Forecasting Methodology',(25) it is stated,
'The first requirement in the forecasting process is to establish a
view about the growth of the economy over the forecast period'.

Here is how the requirement is met by the Department of Energy:
> The view has been taken that a reasonably successful
> economic policy would result in an average rate of growth
> in gross domestic product of 3 per cent over the forecast
> period, that is broadly in line with the growth of productive
> potential. An alternative scenario is also usually considered
> in which the average rate of growth of GDP is put at around
> 2 per cent, although this if sustained would be likely to lead
> to steadily rising unemployment. It will be noted that they
> [the growth assumptions] are not crudely exponential but
> postulate some slowing down in growth in the later years
> under the influence of demographic and other changes.

The assumptions about economic growth are the foundation of
the energy forecasts and this is all that 'Energy Forecasting
Methodology' has to say about them. If we seek further enlighten-
ment in the Department of Energy's discussion paper 'Energy
Policy' (26) we find that discussion of the growth assumption is
again confined to a single paragraph in which it is made clear
that the alternative 'low' growth rate assumptions are not to be
taken seriously:
> The main variant considered to a 3 per cent GDP growth
> rate is of continued sluggishness in our economic growth
> performance, slowing down to under 2 per cent by the end
> of the century. This case,which would imply major problems
> of economic adjustment, not least in unemployment, is not in
> itself seen as particularly credible but is a means of estab-
> lishing a generally lower demand for energy.

The assurance, in 'Energy Forecasting Methodology', that the
forecasters have taken account of demographic and other changes
beyond the end of the century, is particularly intriguing. Nothing
more is said about the 'other' changes, but the taking account of
demographic changes by shaving small amounts off the growth rate
thirty years hence can only be based on population forecasts.
Figure 10.6 describing the recent history of population forecasting
in Britain, indicates the sort of information upon which the shaving must have been based.

Table 10.1 lists some of the GDP growth assumptions in recent
or current use for forecasting purposes by government departments. No explanation is offered in any of the sources of Table
10.1 for the variety in the assumptions about this most fundamental of inputs to the forecasting models. While the difference may
seem slight, when compounded over a thirty-year period they produce very substantial differences. For example, after thirty years
the *difference* between a GDP that has grown at the lowest rate in
Table 10.1 and one that has grown at the highest rate is almost
two times greater than the GDP at the beginning of the period.
There is also disagreement among the Departments about whether
the rate of economic growth will increase or decrease. The Departments of Transport and Energy expect it to slow down while
the Department of Trade expects it to be higher at the end of the
forecast period than at the beginning. Given the long lead times

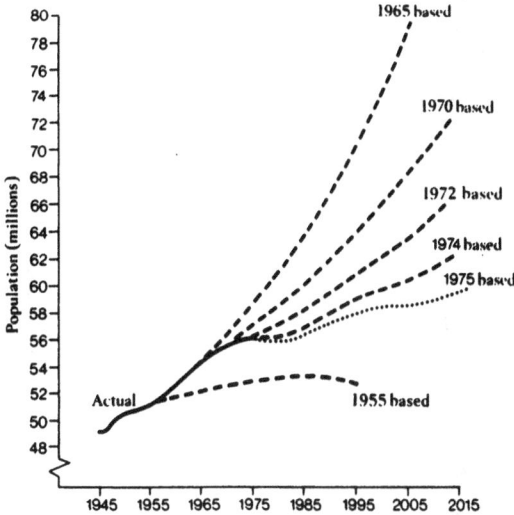

Figure 10.6 Past projections of population
Source:  Department of Transport (1977), 'Report of the Advisory Committee on Trunk Road Assessment', HMSO.

required for major transport and energy projects this uncertainty
has great significance for the present. In its most recent memorandum on forecasting, the Department of Transport describes the
period between 1995–2005 as 'the critical period for which roads
are currently being designed'.(27) And in 'Energy Policy' it is
pointed out that 'even if we were to decide to press ahead with
the first commercial [fast reactor] station at once, it would be
about the end of the century before fast reactors could be making
a significant contribution to our energy supplies'.

It is also instructive to compare what the Department of Energy
and the Department of Transport have to say about future energy
prices. In turning over to the Department of Transport responsibility for the traffic forecasts that are such an important element
of its own forecasts, the Department of Energy observes, 'So far
as road traffic is concerned the critical assumptions which have
to be made concern the pump price of petrol (a function of tax-
ation as well as the cost of crude oil) and the average efficiency
of the vehicles'.(28) With the exception of the trivial allowance
made for variations in car use discussed above, the Department of
Transport's most recent forecasts treat these critical assumptions
as being of no significance whatsoever. This perhaps is just as
well since a glance at the range of official assumptions made in
recent years by the Department of Transport about the price of
petrol, shown in Figure 10.5, suggests that the Department does
not have the slightest idea what to use for assumptions; the high

*Table 10.1   Economic growth rate assumptions*

|  |  | Department of Trade | | Department of Energy | Department of Environ-ment+ | Department of Transport |
|---|---|---|---|---|---|---|
|  |  | (a) | (b) |  |  |  |
| 1975-80 | low | 1.2 | 2.0 | * | 2.5 | 3.0 |
|  | high | 2.4 | 3.0 | 3.0 | 3.5 | 4.0 |
| 1980-85 | low | 2.7 | 2.5 | * | 2.5 | 3.0 |
|  | high | 3.7 | 4.5 | 3.0 | 3.5 | 4.0 |
| 1985-90 | low | 2.2 | 2.5 | * | 2.5 | 2.0 |
|  | high | 3.2 | 3.5 | 3.0 | 3.5 | 3.0 |
| 1990- | low | n.a. | n.a. | * | 2.5 | 2.0 |
|  | high | n.a. | n.a. | ** | 3.5 | 3.0 |

NB The Department of Trade, for purposes of forecasting air traffic, uses two economic indices. (a) is per capita consumption and is used for forecasting leisure traffic, (b) is GDP and is used for forecasting business traffic.

* 'slowing down to under 2 per cent by the end of the century'.
+ rate is assumed to continue 'in perpetuity'.
** The GDP growth assumptions 'are not crudely exponential but postulate some slowing down in growth in the later years . . .'

Sources: Department of Trade, 'United Kingdom Air Traffic Forecasting: Research and Revised Forecasts', 1978; Department of Energy, 'Energy Policy: a Consultative Document', 1978; 'Department of Environment, Standard Forecasts of Vehicles and Traffic', Technical Memorandum H3/75, 1975; Department of Transport, 'National Traffic Forecasts', Interim Memorandum, 1978.

price assumption in use up until early 1978 is now less than the revised low price assumption.

## TRANSPORT POLICY: CHICKEN OR EGG?

Figure 10.7 illustrates for the pre-1978 road traffic forecasts the relationships between assumptions, policies and forecasts. Although the detail of the forecasting models differs over time and varies according to the type of traffic being forecast, the circularities illustrated have persisted.

The forecasts of road or air traffic in a future year, $t_x$, are derived from equations that treat traffic as a mathematical function of a set of assumptions. Policy is to build the road or airport capacity necessary to cope with the traffic forecasts. An absolutely crucial assumption made by the forecasters is that this capacity will be built. Sometimes this assumption enters directly and explicitly into the forecasting calculations. For the road traffic forecasts this assumption is illustrated by the solid line in Figure 10.7 linking the box containing the policy statement (upper right-hand corner) to the equation used to estimate the average number of miles travelled per car. The greater the number of miles of motorway that the forecaster assumes will be built, the greater the volume of traffic he will forecast, and the greater the number of miles of motorway will be built.

# ROADS POLICY:
# CHICKEN OR EGG?

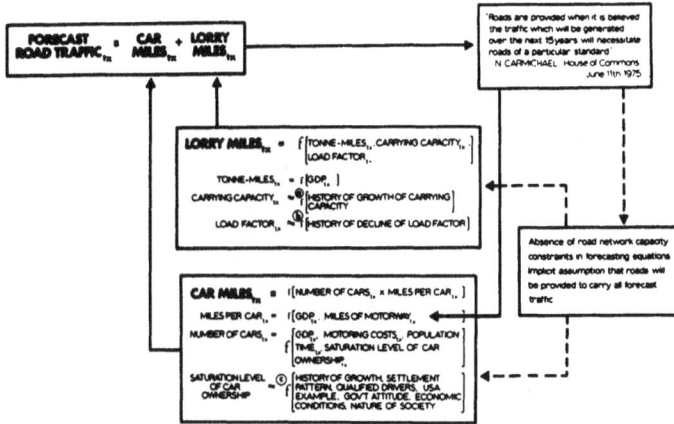

Figure 10.7 Roads policy: chicken or egg?

Because the number of miles per car appears not to be highly sensitive to the number of miles of motorway, this circularity is not very important. There is a much more important implicit circularity in both the road and air transport planning procedures: none of the road traffic or air traffic forecasting equations contain any constraints on the growth processes that are being forecast. This is illustrated for the case of road traffic forecasting by the broken line on Figure 10.7. Thus, as noted above, the environment of the growth process is assumed to be free and unlimited.

'In making our forecasts,' explains Mr Tanner, 'we have tried
to say what we think is likely to happen, not to say what is desirable.'(29) It is the job of those formulating policy to say what
they think is desirable. It is not an easy task. People disagree.
But the government, in declaring its policy to be the provision
of roads and airports to cope with the forecast traffic, has virtually abandoned its responsibility for policy to the prophets -
and the prophets have declined to accept it.

REFERENCES AND NOTES

1  Department of Trade (1978), 'Airports Policy', HMSO, Cmnd
   7084.
2  Tanner, J. C. (1977), 'Car Ownership Trends and Forecasts',
   Transport and Road Research Laboratory, LR799.
3  Department of Transport (1977), 'Transport Policy', HMSO.
4  Tanner, J. C. (1978), Long-term Forecasting of Vehicle
   Ownership and Road Traffic, 'Journal of the Royal Statistical
   Society', 141, Part 1, pp. 14-63.
5  'Hansard', 28 April 1978, col. 1945.
6  Morrell, J. (ed.) (1977), '2002: Britain Plus 25', The Henley
   Centre for Forecasting.
7  Department of Transport (1977), 'Report of the Advisory Committee on Trunk Road Assessment', HMSO, p. 111.
8  Tanner, J. C. (1975), 'Forecasts of Vehicles and Traffic in
   Great Britain: 1974 Revision', Transport and Road Research
   Laboratory, LR650.
9  Department of the Environment (1975), 'Standard Forecasts
   of Vehicles and Traffic', Technical Memorandum H 3/75.
10  See, for example, the section entitled 'Absurd scenarios' in
   Chapter 13.
11  Chapter 9, reference 14. These 'interim' forecasts were published early in 1978. Two years later, even though the forecasts were being used regularly at public inquiries, the
   details of the method by which they were derived had still
   not been published.
12  Tanner (1975), op. cit.
13  Ibid.
14  This incident is discussed in Chapter 11.
15  Department of Energy (1978), 'Energy Forecasting Methodology', Energy Paper Number 29, HMSO.
16  Department of Trade (1975), 'Airport Strategy for Great
   Britain, Part 1: The London Area', A Consultation Document,
   HMSO.
17  Department of Trade (1978), 'United Kingdom Air Traffic
   Forecasting: Research and Revised Forecasts', HMSO.
18  Perhaps the most noteworthy of the assumptions made about
   costs is that relating to the price of fuel. The 'pessimistic'
   low growth forecast assumes an increase in fuel prices of 50
   per cent between 1978 and 1990. By the end of 1980 this

figure had already been exceeded by a wide margin. The forecasters' pessimistic scenario thus assumes a substantial decrease in fuel prices in the 1980s.

19  Commission on the Third London Airport (1970), 'Papers and Proceedings', HMSO, vol. VII, p. 16.
20  Civil Aviation Authority (1975), 'Airport Development in South Wales and South West Region of England', CAA Cap 377/75.
21  Kahn, H. (1976), 'The Next 200 Years', William Morrow & Co., p. 40.
22  Meacher, M. (1978), Tourism Is Good for You, 'Sunday Times', 27 August.
23  Department of the Environment (1975), op. cit.
24  Department of Transport (1978), 'National Traffic Forecasts', Interim Memorandum.
25  Department of Energy (1978), 'Energy Policy: a Consultative Document', HMSO.
26  Ibid.
27  Department of Transport (1978), op. cit.
28  Department of Energy (1978), 'Energy Forecasting Methodology', Energy Paper No. 29, HMSO.
29  Tanner, J. C. (1975), Letter to 'Town and Country Planning', February.

# 11 ASSESSING

Respectable secretary with respectable firm is sent to
collect documents from West End hotel; is there accosted
by young Arab with offer of £500 for the night, instantly
increased to £1000 in face of indignation. Indignation sub-
siding upon reflection that it would take three years to
save so much, she accepts. Consternation next morning
upon finding herself alone, then she notices a cheque for
considerably more than agreed, indeed an acceptable
deposit for the flat she wants. She works on contentedly
as a respectable secretary with a respectable firm.
A tale of our times reported in 'Evening Standard',
17 November 1977.

The Advisory Committee on Trunk Road Assessment appointed by
the Secretary of State for Transport to inquire into criticisms of
the Department's assessment methods concluded that the methods
were not comprehensible to the general public, and did not command its respect.(1) The problem remains.

Before the Advisory Committee's review, the Department's formal quantitative procedures for assessing road schemes consisted
of a cost-benefit analysis programme called COBA plus a set of
methods for measuring the noisiness and visual intrusiveness of
alternative road schemes. In addition, inspectors at public inquiries commonly heard and commented upon evidence of a less
formal nature dealing with opinion about the consequences of the
proposed schemes for the local area.

The Advisory Committee's principal criticism of this practice
was that it was overly dominated by COBA. It had no fundamental
criticism to make of either COBA or the Department's methods for
mapping the audio-visual consequences of its road schemes. Indeed, it stressed that in its opinion these methods were 'basically
sound'. The Committee's remedy for the fault in emphasis that it
had found was a 'framework, relying on judgment, which embraces
all the factors involved in scheme assessment'.

This framework turns out to be a cost-benefit sandwich. It
begins with a summary of road user benefits taken, with minor
modification, from COBA, and concludes with a summary of the
road construction and maintenance costs, also taken from COBA.
The filling of the sandwich consists of a list for the local area, of
the following: the audio-visual effects of the scheme, the numbers
of buildings to be demolished, the amount of land to be taken, the
number of farms to be 'severed', and the number of jobs to be
lost or gained, plus a ranking of the schemes being considered in

terms of the favour they find with various levels of local government, their impact on local parks, conservation areas, historic buildings, archeological sites, and something called 'ecological value'. Also between the costs and benefits in the framework are boxes to be filled with verbal descriptions of the effects on the landscape, historic buildings, archeological sites, local resources, other forms of transport and, again, ecological value. The filling contains no factors that have not previously been discussed in inspectors' reports. It is simply a check-list of factors that the Committee thought should be considered in the assessments of all trunk road schemes.

## COST-BENEFIT ANALYSIS

Although the Advisory Committee recommended that COBA's domination of the assessment procedure should be reduced, it thought that COBA itself should be retained more or less intact. In defence of COBA the Committee argued that the size and complexity of the trunk road programme (it contained at that time over 400 schemes under active consideration) made it necessary to have a standardized evaluation procedure to ensure consistency and to preserve central control of the decision-making process.

Consistency and control are purchased at the price of arbitrariness. COBA embraces but a tiny, arbitrary fraction of the costs and benefits of a major road scheme. The only benefits it considers are savings to road users of time, vehicle operating costs, and accidents, and the only costs it deals with are the costs of building and maintaining the road. Cash valuation of these four elements yields a quite arbitrary cost-benefit ratio. Of the four, one is a hard cash element (construction and maintenance costs), one is generally insignificant (vehicle operating costs),(2) one is highly contentious (time savings),(3) and one is meaningless (the money cost of accidents).

The logical knots that cost-benefit analysts tie themselves into when trying to calculate the cash value of people killed in accidents are discussed at some length in Appendix I, but it is worth noting here a difficulty that arises when there is uncertainty about whether the net effect of a project, in terms of fatal accidents, is a benefit or a cost. If a project yields a benefit in terms of lives saved then it is reasonable to ask how much people might be prepared to pay for it. Commonly the answer is little or nothing. This is because life is risky and death is certain. The number of 'natural' risks to which we are exposed is very large relative to the protection we can afford to buy with money, and at the most we can only hope to buy a deferment of the inevitable. A common response to invitations to buy small extra bits of protection is a fatalistic shrug. But if a project threatens to inflict death or serious injury then the problem is different. In this case, if the cost-benefit game is being played honestly, according to Pareto's rules, the question that must be asked is how much money

would people demand to be paid before they would willingly allow
the threat to be imposed upon them.(4) The answer to this question
is often a very large sum indeed, and if the risk is seen to be
great, the answer commonly is a sum that exceeds all the money in
the world. The most that can be said for the figure of £42 000
(1976 prices) currently used by the Department of Transport's
road scheme assessors is that it falls within the range of zero to
infinity commonly uncovered by empirical research (see Appendix
I). Unfortunately, in its application to the assessment of schemes
that are part of a larger programme to promote and accommodate
the growth of traffic, the figure is frequently given the wrong
mathematical sign. If the traffic that is promoted by a policy
results in more accidents than are saved by relieving accident and
congestion blackspots, then the accident item in the cost-benefit
analysis ought to be treated as a net cost, not a benefit as in
COBA.

An important source of the confusion about what is a cost and what
is a benefit is the highly restricted geographical scope of COBA.
By relieving roads that are currently dangerously congested, new
road schemes will, it is argued, save time and scarce fuel, and
prevent accidents. But these are highly local benefits that are
more than dissipated over the rest of the country's road system.
Since the war many billions of pounds have been spent building,
improving and maintaining Britain's roads. This expenditure deserves a large share of the credit for the traffic growth that has
occurred since the war. It has both generated traffic directly, by
saving motorists billions of pounds worth of time (this assumes that
most of the expenditure would show a positive return if tested by
COBA), and indirectly by removing road capacity constraints on
growth. Associated with this growth in traffic has been an increase in time spent in cars, in fuel consumed and in numbers
killed and injured on the roads.

As Figure 7.7 on p. 112 shows, the statistical relationship between road building and serious road accidents provides no support for the view that road building saves lives. Because of its
emphasis on the removal of traffic bottlenecks the road building
programme has facilitated a pervasive increase in traffic and,
thereby, traffic accidents.

The time-saving benefits attributed to new roads in the Department's cost-benefit analysis model are also illusory. If time is
money, money is also time. Between 1946 and 1976 the money
spent on buying and running cars, plus the money spent on public transport, increased from 6 per cent of total consumer expenditure to 13 per cent.(5) Thus the proportion of working time
devoted to earning the money to pay for travel has increased very
substantially, and since it is extremely unlikely that the large
increase in the number of passenger-kilometres since the war has
been accompanied by an equivalent increase in the average velocity
of travel, the actual time spent travelling must have also increased
substantially.(6) We seem to be on a treadmill, running faster and
faster, but still falling behind. And it is our running that makes
the speed of the treadmill increase.

Of the classifications of time used in COBA the most important
is 'working time'; over half the benefits of a typical road scheme
are attributed to savings in working time. Working time is considered more valuable than non-working time because it is productive. Savings in working time increase productivity and make
the economy grow. And economic growth increases the benefit-
cost ratio of road schemes in three ways. First, the strong historical correlation between gross domestic product and traffic is
built into the traffic forecasting model in such a way as to transform assumed future economic growth into future traffic growth.
And as traffic grows, so do the road user benefits in the COBA
model because the assumed time saving, accident cost saving and
operating cost saving for each vehicle using a scheme are multiplied
by the number of vehicles expected to use it. Secondly, because
rich people are prepared to spend more money on saving their
lives and their time than poor people, both the value of time and
the value of life used to calculate the benefits of a scheme are
assumed to increase as GDP increases. The two effects amplify
each other. According to the 'optimistic' forecasts of the Interim
Memorandum (see Chapter 9), traffic will increase by 84 per cent
by 2005 and per capita GDP by 140.4 per cent. These increases
combine to produce an increase in 'benefits' of 342 per cent.
Thirdly, the future increases assumed in the value of time and
life are stated to be 'real' increases. These increases reflect an
assumed increase in the productivity of the work-force over time.
Thus 'real' construction costs ought, according to this assumption,
to decrease over time.

The road building bias of the model can be seen more clearly
the further one projects its use into the future. The value of the
time and lives saved by paving extra bits of the country increases
exponentially while the costs of the paving change in a similar but
opposite fashion. If COBA is used for long enough it will justify
paving everything – by which time we will all be very wealthy.(7)

## THE NON-ECONOMIC COMPONENT OF THE ASSESSMENT

Although it was impressed by the need for consistency and by the
importance of reducing value judgments to an absolute minimum,
the Advisory Committee felt that it was 'unsatisfactory that the
assessment should be so dominated [as it previously had been] by
those factors which are susceptible to valuation in money terms...'
Thus, although it runs directly counter to the grain of the arguments for consistency and centralized control that it deployed in
its defence of cost-benefit analysis, the Committee recommended
expanding the area of judgment in the assessment process. Here,
in the filling of the COBA sandwich, is where one might hope to
find checks against the road building proclivities of COBA described above.

Such hopes are vain. The judgmental part of the assessment
fails utterly to live up to the promise of dealing with all the

important non-economic consequences of building new roads. It
dwells at inordinate length on trivia and ignores what is most
important. This part of the assessment is concerned with minimizing a few fairly minor consequences of traffic growth, and not
with a consideration for the desirability of the growth. Growth's
beneficial consequences are simply assumed.

The non-economic part of the assessment is made up of a mixture
of factors, some of which are merely listed and described verbally,
such as historic buildings, and others that are quantified, albeit
not in cash. If one judges the relative importance that the Committee attached to these factors in terms of the labour required to
reduce them to the form in which they finally appear in the
assessment framework, then it appears that noise and visual intrusion are considered of overriding importance. As with the
economic component, a consideration of these factors is restricted
to the local area. Far from checking the tendencies of COBA, the
assessment of these factors is constrained in such a way as to
reinforce them. If the growth of traffic is assumed, then a 'do
nothing' solution that would allow it to pile up on existing congested routes through residential neighbourhoods and shopping
streets would obviously result in it being seen and heard by more
people than if it were diverted to bypasses through less densely
inhabited areas. The bypass alternatives that save the most time
by relieving congestion also tend to achieve the highest benefit-
cost ratios. In 'Noise in the Next Ten Years' the Noise Advisory
Council estimated that traffic growth trends would cause a threefold increase between 1970 and 1980 in the number of people in
Britain regularly exposed to noise levels above 70dBA.(8) Similar
estimates have not been made for the visual intrusion effects, but
since they are held to be positively related to the volume of traffic,
they too can be expected to increase nationally. Thus any local
reduction in noise or visual intrusion produced by the government's road schemes are expected to be overwhelmed by the pervasive increases produced by government traffic promotion policies.

The Advisory Committee recommends that the assessment of
noise be based on noise contour maps, and the assessment of
visual intrusion on something similar called visual envelope maps.
These permit various alternative schemes to be described in terms
of their effect on the numbers of people who will hear traffic noise
above a given level, and who will be able to see traffic and traffic-
carrying structures. These two factors present valuation problems every bit as intractable as those encountered in the economic
part of the analysis. The Committee shies away from attaching
cash values to them, although why it thinks that life is 'susceptible to valuation in money terms' but peace, quite and beauty are
not, it does not say. In the end it resorts to crude head counting,
or sometimes dwelling counting. It assumes that the scheme which
permits the traffic to be seen and heard from the smallest number
of dwellings will be, audio-visually speaking, the best.

In the assessment of the visual consequences of a scheme, vis-
ibility is equated with visual intrusion, and visual intrusion is

considered undesirable; the method refers not to dwellings from which roads and traffic will be highly visible but to dwellings subject to 'severe visual intrusion'. The method avoids considering whether a scheme and its traffic consequences will be beautiful or ugly by simply assuming that the fewer the people who can see it, the better. It is a method to be employed, ironically, by engineering firms whose offices and reception rooms have walls proudly covered with pictures of their award-winning road building achievements.(9) Obviously roads that some people think beautiful, other people think ugly. Such differences of opinion are often at their most acute when the road under consideration is a contentious one. For example, a person is unlikely to judge beautiful, no matter how artfully landscaped, a road that will cut him off from his neighbours. Nor is a person likely to consider ugly a view of traffic bearing customers to his place of business. Assessing the visual appeal of a road scheme is not the same as judging abstract paintings in an art competition. In trying to decide whether or not they like the look of a scheme, people cannot separate matters such as the texture and composition of a view from the question of how it will affect them. Nor should they be asked to try.

The Advisory Committee's method of assessing noise favours the scheme that will do most to encourage the growth of traffic while at the same time doing the most to suppress the acoustic evidence of growth. Where no route for a scheme can be found that will not inflict noise above 68dBA on some dwellings, the dwellings are provided with free ear plugs in the form of double glazing. It seems an inconsistency in the Department of Transport's methods for suppressing the evidence of traffic growth that people who do not like the sight of traffic are not offered tinted double glazing.

In brief the charge against the present methods of road scheme appraisal, even with the modifications recommended by the Leitch Committee, is that they deal with a relatively insignificant proportion of the consequences of proposed new roads, and deal with it arbitrarily.


AN INSIDIOUS POISON

In its 1976 discussion paper 'Transport Policy', the Department of Transport conceded a number of major criticisms of cost-benefit analysis:
> Not everything can be given a satisfactory proxy money value. Environmental factors such as noise, vibration, air pollution, visual intrusion and community severance can be measured physically, and are taken into account as far as possible in reaching decisions, even though, as yet, it has not proved possible satisfactorily to value these effects in money terms, so as to use them on a regular basis for evaluation. (II.5.4)
> and: At present such effects [the land use consequences of road proposals] cannot be taken into account systematically

in a quantitative way, since this would be an extremely
arduous and uncertain task.... For all these reasons these
factors, together with many others, have to be judged at
planning inquiries. (II.5.26)

The words 'as yet' and 'at present' hold out the hope that the
scope of the Department's systematic quantitative methods can be
extended and their arbitrariness reduced. The most common defence of cost-benefit analysis to the sort of criticism that has been
made above consists of confessing present lack of perfection and
urging bigger and better cost-benefit analyses for the future.
The conclusion to Professor Williams article, Cost-Benefit Analyses: Bastard Science? And/Or Insidious Poison in the Body
Politik?, exemplifies this defence:

The ... great danger is that the perfect becomes the enemy
of the good, that the acknowledged limitations of the new
product are made the excuse for not abandoning old practices
which have even more defects, on the curious notion that we
should only change over if a perfect product is offered in
place of the imperfect one we are already using.(10)

Although the Department of Transport appears to be developing
reservations about the feasibility of comprehensive cost-benefit
analysis it remains an article of faith among proponents of cost-
benefit analysis, Layard (11) reminds us, that 'rational choice'
requires the reduction of all aspects of a choice to a common denominator. 'The basic decision rule' of cost-benefit analysis,
Pearce (12) asserts, 'requires that benefits and costs be expressed in monetary units'. Unless valuations are expressed in money
terms, Williams argues they are 'vacuous'.

While admitting that their science is not yet at the stage where
this cash valuation can be done with any degree of precision this
is the acknowleged imperative of the work now being done on the
weights that ought to be given to 'environmental factors'. Hopkinson,(13) for example, discussing his work on visual intrusion
insists 'Amenity has to be costed in relation to other requirements'
and the TRRL working group on research into road traffic noise
speaks of the need for 'intensive research ... both on the costs of
noise reduction and on the [economic] evaluation of the amenity
gained'.(14) Work now being undertaken with the TRRL's 'environmental simulator' is also aimed at establishing monetary values for
changes in the audio-visual environment produced by roads and
traffic.(15)

Currently cost-benefit analysis is beset with a host of valuation
problems and as yet it is not possible to make a comprehensive
assessment of the costs and benefits of road proposals; thus at
present cost-benefit analysis can serve only to assist judgment
and not, by itself, to take decisions. The goal, however, while
still distant, is clear. What is aspired to, in the name of 'rationality', is a 'Solomon Machine', into which can be fed accurately
measured valuations of all the factors bearing upon a decision and
out of which will come even-handed justice for all. It is a perspective on planning problems that transforms criticisms of cost-

benefit analysis into arguments for more cost-benefit analysis; the more damning the criticism the more urgent the need, in Professor Williams's words, 'for pressing on as fast as possible with the quantification of the currently unquantifiable'.

The Pareto principle upon which cost-benefit analysis rests requires that for a proposed road to represent a 'Pareto improvement' it must make one or more people better off while leaving no one worse off. This requires the benefits to be such that after all losers have been compensated for their losses, there is still something left over for the winners. This, at least at present, is considered by most practitioners a Utopian criterion: '... there is almost no case where it is feasible to compensate everybody, and if the Pareto rule were applied no projects would ever get done'. (16) Thus, faced with our present imperfect institutions for ensuring compensation, most economists are prepared to settle for a potential Pareto improvement as justification for a project. This requires only that the gains be such that the winners could compensate the losers and still come out ahead. But the Pareto principle, even in its potential form, requires that both individual winners and individual losers be the valuers of their potential gain or loss; any mooted change must be valued in terms of what the winners would be prepared to pay for it and what the losers would consider satisfactory compensation.

The estimation of the compensation payable to losers encounters an insurmountable problem; in the solution to this problem resides the insidious poison. The problem is what to do with the (often large number of) people who insist that the loss that they will suffer if a project goes ahead is simply not compensatable by any amount of money. Since just one such person will cause a project to fail the Pareto improvement test, the test in effect gives every individual a veto, and this makes the test useless for practical planning purposes.

The nature of this limitation is explained clearly and in some detail by Mishan (17) but, presumably because it is such a severe limitation, it is widely disregarded in the literature on cost-benefit analysis. For example, Pearce in his book 'Cost-Benefit Analysis' skips past the problem as though it were of little significance: 'Where the potential external effect is in the form of a cost, the surrogate price could be formulated as what the individual is willing to pay to prevent the nuisance. In practice, the more acceptable measure is what he is willing to accept in compensation to put up with the nuisance.'

Nash, Pearce and Stanley, in their review of cost-benefit analysis, (18) also slide round the problem by advocating the measurement of the effects of a proposal 'in terms of what individuals would be willing to pay for them if they all had some specified income level per head'. Similarly Layard in his introduction to 'Cost-Benefit Analysis' says 'the broad principle is that we measure [a person's] change in welfare as he would himself value it; that is we ask what he would be willing to pay to acquire the benefits or to avoid the costs'. The questionnaire administered by the TRRL to try to ascertain the cash value that people place on environments threatened by roads asks 'How much would you pay annually to prevent your own environment changing to the next worst situation?' - the next worst situation being a scene displayed on a screen in the Laboratory's environmental simulator.(19)

None of these examples betrays any awareness of the absolutely vital distinction, discussed above with reference to the value of life, between the price that someone is prepared to pay to safeguard something that he values, and the price that he would consider fair compensation for its loss. The former is frequently a, literally, infinitessimal fraction of the latter, and to use the former in place of the latter is to abandon the very essence of the Pareto principle. In practice, and in theory, the only acceptable measure of loss is what someone is willing to accept in compensation.

It is not, as Pearce implies, a matter of little consequence, almost a matter of taste, whether one prefers, in the jargon of economics, a 'compensating variation' to an 'equivalent variation'. And it is not merely a problem of some people being willing to spend more money than others to protect what they value because they have more money. Even in an egalitarian society of millionaires, the gulf between the two measures would remain unbridgeable - perhaps even wider because millionaires could be expected to be the most aware of all the things of value that money cannot buy.

The Roskill cost-benefit analysis did recognize the distinction between these measures and endeavoured, by means of a questionnaire survey, to find out how much people would agree to accept as compensation for the loss of their homes. Some 8 per cent of those surveyed insisted that no amount of money could compensate them for such a loss (a similar British Airports Authority survey recorded 38 per cent in this category).(20) Such answers, if accepted at face value, should require either abandoning the project or abandoning cost-benefit analysis. Cost-benefit analysts are loath to do either and instead impugn the rationality or the integrity of those who insist that they value some things above monetary compensation. Pearce observes 'These replies [of those who could not be compensated by money] would seem to be inconsistent with the general view that "each man has his price". If the response is ascribed to some element of irrationality in the householder, the problem arises of how to treat the factor in the cost-benefit analysis. The procedure in the [Roskill] study was to truncate the distribution at some arbitrary level.'(21) The arbitrary level that the Roskill Commission chose to substitute for infinity was 200 per cent above the market value of the property. In puzzling over why so many people claimed that no amount of money would compensate them for the loss of their homes the Commission speculated that perhaps some people did not answer honestly because they did not appreciate the hypothetical nature of the questionnaire and were bargaining for the highest possible price.

The exasperation of the cost-benefit analyst with a man who cannot or will not name his price, illustrates what an insiduously corrupting poison cost-benefit analysis is. It used to be a common view that people ought to hold certain things, the most valuable things, above price. The extent to which this view is less common than previously is a measure of the increased acceptance of the cost-benefit ethic. It is an ethic that debases that which is important and disregards entirely that which is supremely important.

## THE WISDOM OF SOLOMON

Cost-benefit analysis is a nonsense. It stands even less chance of reducing the basic elements of transport planning decisions to gold than did the rituals of the medieval alchemists who attempted to perform a similar feat with rather more tangible base elements. Science has been called the art of the soluble; cost-benefit analysis is one of the black arts of the insoluble.

The economist's search for a comprehensive, systematic, consistent, quantified, and centrally controllable decision-making method assumes that large and complex social issues can be broken up into their constituent parts, that all the parts can be weighed in the banker's scales and then reassembled to provide a decision. It is a search for a Solomon Machine, a machine that will embody in quantified form the principles of profit maximization and distributive justice, a machine into which can be fed accurately measured valuations of all the relevant facts, and out of which will flow wealth and even-handed justice for all.

Solomon decided to chop the baby in half. If the claims of two or more people to something of value appear to be of equal merit, then dividing this something into equal portions is a solution that in terms of equity has much to be said for it. Indeed if the thing contested is, say, a pot of money, few could find fault with the justice of such a solution. Cost-benefit analysis chops a problem into a multitude of pieces, some good, some bad, 'values' each one and then tries to ensure that the value of the good exceeds the value of the bad. There is also usually, although not always, an attempt made to ensure that the good and the bad are shared out equitably. But contemporary transport planners, like Solomon, are dealing with the fate of living entities. Half a baby is murder. Cost-benefit analysis is vivisection.

The wisdom of Solomon's decision begins where cost-benefit analysis leaves off. Solomon realized, in the dispute about the baby, that he was confronted by two irreconcilable claims of unequal merit; and this is the nature of most disputes about transport schemes. Both sides to such disputes claim a vital interest. One side claims that growing traffic is the life blood of a healthy economy, the other that it is literally and figuratively killing society. One side argues for traffic accommodation and more traffic, the other for traffic restraint and less traffic. The two policies aim in opposite directions. It is not a dispute that lends itself to a chop and share solution.

In Ontario, the Ministry of Transportation and Communications used to include in its list of things that need to be valued in money terms an item called 'community cohesion'.(22) A community's transport system serves as both its nervous and circulatory systems combined. It is literally vital to its existence; human communities are distinguished from mere collections of individuals by the coherence of their internal social relations. There are countless communities whose cohesion has been destroyed by increases, or decreases, in the mobility of ther members. Whether the new social relations that might evolve would be an improvement on what might be lost as a consequence of transport developments is the question that ought to be discussed and debated. It is not a question to be found in the Leitch Committee's framework. It is a question that embraces our feelings about our friends and neighbours, and the sort of society in which we wish to live. And it is a question that is grotesquely debased by attempts at cash quantification.

The Roskill Commission (23) insisted that its method of analysis required the assumption that 'every man has his price', because in a comprehensive cost-benefit study all the significant consequences of a scheme must be reduced to cash. The pursuit of every man's price is corrupting. At the time of the M3 inquiry at Winchester, Mr Thorn, the headmaster of Winchester College, spoke of the 'desecration' threatened by the 'criminally wrong' proposals of the Department of Transport.(24) Another word commonly used to describe the government's transport projects is 'rape'; Sir Colin Buchanan (25) called the anticipated consequences of a major new London airport 'a rape of the English countryside'. To try to persuade people that they ought to consider how much money would induce them to change their minds about the sacredness of their community or the inviolability of the English countryside is an attempt to corrupt them. Rape, preceded by cash compensation willingly accepted, is difficult to distinguish from prostitution.


REFERENCES AND NOTES

1   Department of Transport (1977), 'Report of the Advisory Committee on Trunk Road Assessment', HMSO.
2   Department of the Environment (1976), 'Transport Policy: A Consultation Document', HMSO, vol. 2, p. 99.
3   See, for example, Harrison, A. J. and Quarmby, D. A. (1969), The Value of Time, in 'Cost Benefit Analysis', R. Layard (ed.) Penguin, 1972.
4   The central rule states that any change that makes at least one person better off while leaving no one worse off is an improvement. For this to be of any practical value to decision-makers the costs and benefits of the change must be measurable in common units. Pareto, who is credited with having laid the ground rules for cost-benefit analysis, was hailed by Mussolini

as 'the father of Fascist theory' ('Twentieth Century Authors', p. 1074).

5  Central Statistical Office, 'National Income and Expenditure', 1966-76 and 1946-51, HMSO.

6  Tanner, J. C. (1979), in 'Expenditure of Time and Money on Travel', Transport and Road Research Laboratory Supplementary Report 466, Table 10, estimates that the time spent travelling by the 'average person' has increased by 16 per cent between 1953 and 1976.

7  There is a possibility that the rising price of land might cause the costs to overtake the benefits before this ultimate state is reached. As the supply of money increases over time and the supply of land available for roads decreases, the price of land, and all prices related to it, will increase, eventually very dramatically. Whether these increases are 'real' or inflationary or both seems to depend on which economist one asks.

8  Noise Advisory Council (1974), 'Noise in the Next Ten Years', HMSO.

9  This pride of achievement is exemplified by Sir James Drake's book 'Motorways' (Faber & Faber, 1969): 'The building of a motorway is sculpture on an exciting grand scale...' p. 209.

10  Williams, A. (1973), Cost-benefit Analysis: Bastard Science? And/Or Insidious Poison in the Body Politik, in 'Cost Benefit and Cost Effectiveness', J. N. Wolfe (ed.), Allen & Unwin, London.

11  Layard, R. (1972), 'Cost-Benefit Analysis', Penguin, Harmondsworth.

12  Pearce, D. W. (1971), 'Cost-Benefit Analysis', Macmillan, London.

13  Hopkinson, R. G. (1972), The Evaluation of Visual Intrusion in Transport Situations, 'Traffic Engineering and Control', December.

14  Transport and Road Research Laboratory Working Group (1970), 'A Review of Road Traffic Noise', LR357, Crowthorne, Berks.

15  Dawson, R. F. F. (1974), 'Environmental Simulator: Progress Report', LR659, Transport and Road Research Laboratory, Crowthorne, Berks.

16  Layard (1972), op. cit.

17  Mishan, E. J. (1971), 'Cost-Benefit Analysis', Unwin University Books, London, Chapter 18.

18  Nash, C., Pearce, D. and Stanley, J. (1975), An Evaluation of Cost-Benefit Analysis Criteria, 'Scottish Journal of Political Economy', vol. xxii, 2, pp. 121-34.

19  Dawson (1974), op. cit.

20  Commission on the Third London Airport (1971), 'Report', HMSO, p. 275.

21  Pearce (1971), op. cit.

22  Ministry of Transportation and Communications, 'Priority Analysis', Systems Manual, 1974.

23   Commission on the Third London Airport (1971), 'Papers and Proceedings', HMSO, vol. VII.
24   Thorn, J. (1976), letter to 'The Times', 13 July.
25   Buchanan, C. (1971), Comment following Adams, J. G. U., London's Third Airport, 'Geographical Journal', 1972, p. 501.

# 12  INQUIRING

The 'Report on the Review of Highway Inquiry Procedures' published in 1978 accepted that there was a need to restore the confidence of 'responsible people of moderate views' in the 'fairness of the highway inquiries system'.(1) The Department of Transport has attempted to do this by cultivating the idea that its inquiries are 'independent'.

By why should the Department of Transport ever hold an independent inquiry into anything? Why should it willingly allow persons independent of it to have a hand in decisions about important matters that lie within its own jurisdiction? The answer appears to be that it never does, except through miscalculation or indifference.

'Independent' inquiries are held frequently. They are used by the Department to overcome political obstacles to its policies and to extricate it from controversy with a minimum of embarrassment. Some obstacles and controversies, such as those associated with road building proposals, occur almost as a matter of routine, and the statutory procedures for dealing with them are usually equally routine. Others, such as those that have recently arisen over traffic forecasting, the methods of trunk road assessment, and the regulation of the road haulage industry, occur irregularly and the inquiry procedures for dealing with them are more ad hoc in nature.

Although huge volumes of evidence are frequently examined at inquiries this is not their principal purpose. The civil service possesses, or could acquire, the skilled manpower to do the job itself equally thoroughly. Most inquiries are held not because the politicians and civil servants responsible for the relevant controversy do not know their own minds, but to provide support for policies and decisions that have already been made. This is not to say that inspectors or committees of inquiry are never given any scope to exercise their own judgment. Where a controversy relates to a matter of purely local significance, such as a question of route alignment or landscaping, or to a matter of relatively trivial detail - such as the procedures for monitoring vehicle maintenance - then the Secretary of State might well be totally indifferent to the outcome, and happy to allow an 'independent' inspector or committee to shoulder the resentment of those who do not like his decision. He might even positively welcome recommendations from an inquiry challenging the proposals of his civil servants on relatively minor matters in order to foster the inquiry system's image of independence. But the longer a controversy

endures and the more important it is, the less likely it is that the
Secretary of State and his civil servants will be indifferent to its
outcome. And the less indifferent they are, the less likely they
are to entrust the outcome to an independent inquiry.


## PRESENTATIONAL INQUIRIES

The appointment by the Secretary of State for Transport of a
Standing Committee (the Leitch Committee) charged with subjecting its own advice to 'independent' scrutiny is an example of
spurious independence that has already been discussed in the
chapter on forecasting. The 'Report of the Independent Committee
of Inquiry into Road Haulage Operators' Licensing' is another
example. It is addressed to the Secretary of State for Transport
and begins 'We were appointed by you ...'(2)

It is improbable that an astute politician would deliberately
appoint a committee to inquire into an important matter under his
jurisdiction if he had reason to suppose that it might come to conclusions incompatible with his own views. Even less likely is the
possibility that, in seeking the advice of his civil servants about
whom he ought to appoint, he would be given the names of people
known to disagree with established civil service views.

These deductions receive considerable support from the 'Peeler
Memorandum'.(3) This is a note that circulated among senior civil
servants in the Department of Transport towards the end of 1978
about the advantages and disadvantages of holding an inquiry
into the question of raising the maximum permitted weight for
lorries. The memorandum was intended to be confidential, but it
escaped and received considerable publicity. It makes clear that
an inquiry into the question of lorry weights would not be a real
inquiry into a question on which the Department was having difficulty reaching a decision. Its purpose, clearly stated in the
memorandum, would be 'presentational'; it would be to 'do good
to' the 'sadly tarnished public image' of 'the road haulage interests'.

The value of an inquiry, according to the memorandum, is that
it 'offers a way of dealing with the political opposition to a more
rational position on lorry weights', i.e. of raising the maximum
permitted gross weight of a lorry from 32 tonnes to 38 or 40 tonnes.
Perhaps anticipating the objection that it would be dishonest to
call an exercise intended to burnish the image of the road haulage
interests an 'inquiry', the memorandum proffered the reassurance
that such things are done all the time:

> The fact-finding value of most inquiries of this kind would
> in this case (sic) be heavily subsidiary, since most of the
> relevant evidence is of a technical kind and is already
> available in the Department and the TRRL. However, in
> this respect the proposed inquiry may not be entirely
> exceptional.

The memorandum's view of how the inquiry would work offers further reassurance:

At the end of the day, recommendations would be made by impartial people of repute who have carefully weighed and sifted the evidence and have come to, one hopes, a sensible conclusion in line with the Department's view.

These two passages, judged by standard literary conventions, would qualify as examples of ungrammatical heavy irony. But it is very difficult reading them in their full context, to decide whether the irony is intended, or whether it is the language of people so habituated to talking with their tongues in their cheeks that they have quite forgotten they are doing it. Ironic jesting originally deployed in an attempt to keep conscience alive in an atmosphere of institutionalized dishonesty can easily degenerate into routine double-think.

Impartial people of repute will come to conclusions in line with the Department's view - one hopes. Pious hope. The memorandum discusses whether the inquiry should be conducted in public by an inspector, possibly assisted by assessors (as at Windscale), or whether it should be conducted by a committee, with the public excluded. It concludes that since 'publicity is central to its purpose it might be advisable to hold some of its sessions in public.' It discusses whether the inquiry should be wide ranging, and hence 'difficult to keep under control', or narrow in scope, and hence 'open to accusations of "rigging" '. It is devoted, from beginning to end, to a discussion of what it refers to as the 'means of getting round the political obstacles to change the lorry weights problem itself [sic]'.

The memorandum circulated among the upper-middle management of the Department of Transport. It discussed the arrangements for the inquiry in a matter-of-fact tone that suggests that arranging inquiries that produce the right answers is pretty routine business for the civil servants to whom the memorandum was directed. It was a discussion about what they ought to do, among people who appeared to have the power to do it. Perhaps this power really resides elsewhere in the Department; in theory of course it resides with the Secretary of State. But so long as the power both to dictate the form and scope of an inquiry and to choose the person or persons conducting it resides with the custodians of the Departmental 'view', the custodians need not 'hope' that an inquiry will reach conclusions in line with their view - they can virtually ensure it.

But ensuring that an inquiry comes to a 'sensible' conclusion is only half the battle, the easy half. For a conclusion to be of any political use it must have the appearance of independence from its government sponsors. Otherwise it would simply look like the government repeating itself. In order to be sure of running what the memorandum calls a 'successful inquiry' one must employ partial people with the appearance of impartiality to do the weighing and sifting and concluding.

The importance of appearances emerges very clearly from the

1975 Rees Jeffreys lecture to the Royal Town Planning Institute.
The lecture was given by Alfred Goldstein, a senior partner in
R. Travers Morgan & Partners, a firm of consulting engineers
often employed by the Department of Transport to prepare road
schemes. It was entitled 'Highways and Community Response' and
addressed itself to the problem of the increasingly adverse response of communities through which the Department was proposing to build roads.(4) He began by explaining why the programme
of road construction should continue: 'Highways are an economic
good.' 'Roads are economically worthwhile because...' 'The sheer
convenience of the car...' and much more in a similar vein. He
then asserted that 'the emphasis given by our leading highway
design teams to environmental and community problems is as great
as anywhere in the world'. Having thus established to his own
satisfaction that the work of the government's road builders and
planners was wholly admirable, he proceeded to examine the problem of growing public dissent as a *public relations problem*.

The 'manner' of the Department's presentation of its case was,
he contended, sometimes 'insufficiently convincing': Department
spokesmen sometimes, unwisely, gave an 'impression' of 'commitment' to the schemes they were presenting; the Department was
sometimes 'reluctant' to provide information about its proposals;
the competence of inspectors sometimes created 'doubt'; Department officials were in need of 'tuition' in the 'stock of techniques'
for being effective witnesses. His conclusion was remarkably similar to that of the 'Peeler Memorandum': 'the aim of a public
inquiry should be to engender confidence; to use the opportunity
not only for its strictly statutory purpose, but also to win friends
and influence people'.

People who question the need for a road scheme are suitable
cases for educational treatment, but their doubts cannot be inquired into; that, he felt, would be an inefficient use of public
time and money:

> I am second to no man in supporting individuals and their
> rights. But roads are being and should continue to be built.
> And as that *is* government policy, do not those that do get
> built call for too much, and too lengthy, ancillary effort
> [i.e. spent at inquiries], some of which might be put to
> better public use? I pose the question, I hope, in a balanced
> and non-partisan way.

## JUDICIOUSLY INDEPENDENT

Since 1975 the problem discussed by Goldstein has become much
worse. But it is still viewed as a public relations problem. The
'Report on the Review of Highway Inquiry Procedures', presented
to Parliament by the Secretaries of State for the Environment and
Transport suggests 'Perhaps above all, [people] seek an assurance that they are getting a fair deal; that what they have to say
is given a fair and impartial hearing; and that proper weight is

given to their views by the Secretaries of State for Transport
and the Environment before a decision is reached.'

But the Report evinces no doubt that people are in fact getting
a proper, fair and impartial deal:

Much of the public concern about the fairness of inquiry
processes centres on the fact that the Inspectors, though
not salaried civil servants, are drawn from a fee-paid panel,
managed by the Department of Environment, and appointed
to individual inquiries by the Secretaries of State for Transport and the Environment, one of whom is the promoter of
the scheme and both of whom make the final decisions on the
proposals considered at the inquiry. Some objectors have
argued that this must inevitably prejudice the Inspector's
independence. This suspicion is unfounded. Highway inquiry
Inspectors are chosen for their qualities of open-mindedness
and sound administrative perception and judgment.

In fact, the appointments are made from a panel of highway
inquiries Inspectors whose names are approved by the Lord
Chancellor; but in order to meet the criticism the Secretaries
of State will in future, in exercising their statutory obli-
gations, ask the Lord Chancellor to nominate a particular
individual considered by him to be suitable for a particular
inquiry.

Having insisted that there is no criticism to meet, the government meets it. Since there is no substance to the criticism there
need be no substance to the changes made to meet it. The changes
are purely for the sake of appearances. The crucial part of the
procedure that ensures that inspectors of 'sound administrative
perception' are selected is not the ultimate act of appointment,
nor even the penultimate act of nomination. Both appointer and
nominator are senior politicians of the same party and might be
assumed to be good political friends, but both are busy men.
There are many appointments to be made and these men cannot
possibly have the time to give most of them more than a very
cursory perusal. Thus, in order to ensure that the inquiry system produces a satisfactorily consistent yield of recommendations
that are 'in line with the Department's view' it is necessary to
have a piece of bureaucratic machinery that will produce a short
list of reliable people from which names can be plucked.

Inquiries made of the Lord Chancellor's Office about how names
come to find themselves on the Lord Chancellor's approved panel
of inspectors and how the Lord Chancellor deems one of them
'suitable for a particular inquiry' are met with the answer that
this information is an 'Official Secret'. At the time of writing, the
official in the Lord Chancellor's Office who deals with such inquiries is the same official who used to administer the Panel Support
Group in the Department of the Environment.(5) Although there
has been a shift in the nominal responsibility for the appointment
of inspectors from the Secretaries of State for Transport and the
Environment to the Lord Chancellor, there appears to have been
no change in the personnel administering this responsibility.

At about the same time that additional responsibility for the
inspectorate was being given to the government's senior law
officer in order to assist the impression that inspectors are impartial, it was being revealed that the law officers were regularly
(twenty-five times in the past three years), and without public
knowledge, interfering in the process by which juries were selected. In 'certain exceptional types of case of public importance'
according to 'guidelines' issued by the Attorney-General the
panel of jurors should be investigated to ensure the exclusion of
persons who 'might be influenced, in arriving at a verdict, by
extreme political, racial or similar convictions'.(6)

The Lord Chancellor one presumes (one can but presume in the
presence of official secrets) is equally vigilant to ensure that
persons with 'extreme convictions' do not find their way on to
his panel of inspectors. One is also left to presume about the
guidelines he provides his staff for recognizing such cases. How,
for example, does he define 'sound administrative perception'?
Might it include a keen appreciation of the extreme annoyance and
administrative inconvenience that would result from the turning
down of road building plans of fifteen years' gestation? What, one
wonders, is the transport planning equivalent of a case of
'exceptional public importance'? Might it be a case such as Archway or Winchester or the Aire Valley in which objectors wish to
inquire into the need for the proposed road scheme, and appear
unlikely to be easily satisfied by a conventional 'presentational'
inquiry? And how, in such a case, does the Lord Chancellor pick
a 'suitable' inspector? The 'Peeler Memorandum' spoke of a
'rational position on lorry weights'; might 'suitability' be equated
with the Department's idea of 'rationality'?

In another, not unrelated sphere, the judiciary and the Department of Transport appear to enjoy a close and friendly working
relationship. During the Archway Road Inquiry a private pros-
ecution was brought against a Department official for punching a
GLC councillor. The Treasury solicitor was appointed to defend
the official, and one of his assistants wrote to the Chief Clerk of
the Highbury Magistrates Court where the case was to be heard
suggesting that, because of the 'highly volatile personalities' of
many objectors at the inquiry who appeared to be associates of
the plaintiffs, and because of the possibility of harassment, a
stipendiary magistrate should be appointed to hear the case. Mr
Tobias Springer was duly appointed, a stipendiary who, along
with the Chief Clerk and his fellow stipendiary, has the unique
distinction of having been the object of a picket of barristers and
solicitors. (The pickets, more than seventy in number, were
objecting to the fact that the court at Highbury Corner had been
turning down applications for legal aid at a rate fourteen times
higher than the court in nearby Hampstead.) Mr Springer, having
heard the case, concluded that there was little doubt that the
official had struck the councillor, but that it was not proved that
'it was so reckless that the law says it is an assault'. Despite the
fact that the Department of Transport can draw upon virtually

unlimited resources of legal aid, the plaintiffs were ordered to pay the Department costs of £400. Mr Springer added that he did not want the costs to be seen as a penalty. When the letter from the Treasury solicitor became public after the case, its author protested in tones of surprised innocence that there was nothing at all unusual in his making suggestions of the sort contained in his letter.(7)

During the Archway Inquiry discipline collapsed because respect for the fairness and impartiality of the inquiry procedure had collapsed. People, and institutions, become litigious when mutual respect breaks down. The Department in attempting to bolster discipline by leaning increasingly heavily on the legal arm of government will do nothing to restore respect for its impartiality. It will simply foster suspicion about the impartiality of the legal system.


INESCAPABLE DEPENDENCE

What the Department now calls 'independent public inquiries' used to be called 'ministerial public inquiries'. Because the minister was too busy to inquire personally into all the matters for which he was responsible, he appointed someone with the relevant experience and expertise who was well versed in his policies to do it for him. The inspector was quite openly acting on behalf of the minister. As hostility to the government's road building plans increased, the idea that inquiries should be independent gained favour with both the road planners and their critics. The critics felt that they would get a more sympathetic hearing if they could put their arguments to someone who was not the minister's man. The planners, although initially opposed to the idea, came to appreciate the advantages of an inquiry system with the appearance of independence.

Transport policy is in a state of transition. The brave vision that has guided transport planning for so long, of a congestion free high-mobility future, in which the benefits of cars and airplanes will be available to all, is now foundering on the rocks of social, environmental and economic costs. It is increasingly appreciated that such a future would be socially polarized, environmentally bleak, and economically unsustainable. But the brave old visionaries are still in charge of transport planning.

The White Paper 'Transport Policy' states: 'For the future we should aim to reduce our absolute dependence on transport, see that transport makes its contribution to the conservation of energy resources, and key transport into our wider policies for the environment and our views of how we want our communities to develop.'

However, it also states: 'Increasingly, people have chosen to use their own transport as ownership of a car has come within their reach, and the increase will go on.'

The current post-Leitch forecasts are based on the assumption that before the increase stops there will be at least twice as many cars as there are now; and current expenditure plans call for the spending of over £600 million a year on building new roads.

What are inspectors or committees of inquiry to do with a policy that pulls in diametrically opposite directions? They can make recommendations that pull in one direction, or the other; or they can make recommendations that are as incoherent as the policy itself. But they are not to question the merits of policy. Policy, they are told, provides the framework within which inquiries are to be held, and policy is the business of Parliament and the government.

For many years the inquiry system has been the servant of the old vision. The inquiry is but one stage of a multi-stage planning process. The things being planned were all seen, quite explicitly, as contributing to the attainment of the vision. Not only the Department of Transport officials promoting schemes, but the barristers and solicitors, the consultant experts, the inspectors, and, on the whole, even the objectors shared the vision; during the Roskill Commission inquiry in the late 1960s, to cite a well-known example, the principal argument of those opposing an airport at Maplin was that its remoteness would inhibit the growth of air traffic. Inquiries were into the best and fairest way of attaining the vision.

Although the old vision is foundering, the inquiry system is still dominated by people committed to it. Inspectors usually have a very considerable investment in it. So too do the senior officials in the Department of Transport with responsibility for the schemes being inquired into. They are often people close to retirement age whose careers have been spent in the service of the vision. They are usually highly competent judges of the merits of schemes designed to accommodate growing traffic, but they are notably uninterested in the question of whether traffic growth ought to be fostered and accommodated. As the evidence against the old vision accumulates they are becoming increasingly defensive. To renounce the old vision would be to renounce the work of a lifetime. The oxymoronic 'independent presentational inquiry' is symptomatic of their defensiveness. Rather than conduct a real inquiry into the objections of those who question the desirability of traffic growth, or heavier lorries, they transform the inquiry into a public relations exercise on behalf of the old vision.

Some people are more articulate than others about their visions of future society, but only the utterly stupid or unimaginative can avoid having them. One's vision colours all the evidence presented to one's senses. One cannot be independent of it, and it is this inescapable dependence that guides the recommendations of inspectors and members of committees of inquiry. Where people share a common vision of the future it might well be possible to devise procedures for assessing planning proposals that are 'independent' in the sense that the assessors have no special personal stake in the alternatives being assessed. But wherever conflict involves visionary ends rather than mere practical means there can be no hope of resolving it by means of independent inquiries.

Major transport projects have long periods of gestation. Ten to fifteen years is common and many are much longer. Virtually all major schemes now coming up to the inquiry stage of the planning process were conceived in the period before the old vision encountered serious challenge. That increasing personal mobility is desirable and that facilities should be built to accommodate it are articles of faith of the old vision. The predicament of those who would object to these articles of faith is like that of an atheist obliged to argue before the Pope the case against the existence of God. Both the atheist and the non-believer in the old verities of transport planning have reason to suppose that their arguments would be considered solely for purposes of formulating rejections of them. In the absence of a common set of articles of faith concerning 'how we want our communities to develop', transport planning and the inquiry system continue to serve that half of policy that favours increasing the country's dependence on road transport. If the planners and inquirers were to be guided by the opposite half of policy, controversy would not cease; the planning system would simply acquire a different, more powerful, set of enemies. Appointing an atheist to judge the arguments of an atheist would, after all, be unlikely to satisfy the Christians. Nor can one look to technical solutions for help. So long as the incoherence of policy persists it is pointless searching for a set of systematic methods by which transport projects can be evaluated. It is self evident that a coherent method of evaluation must rest upon a coherent set of values.

The old vision has permeated the very language in which road inquiries are conducted. At most inquiries the scheme being inquired into is referred to by both the inspector and the representatives of the Department of Transport as 'The Improvement'. The official title of a scheme that was inquired into in north London was 'The Archway Road Improvement'. An objector put it to the inspector that a project that would knock down a large number of trees, shops and homes and drive a motorway through the middle of a community was not necessarily an improvement. He asked to have the title changed on the grounds that it anticipated a verdict on what was supposed to be still an open question. The · inspector was unmoved. The word 'improvement', he said, had become a 'colloquialism'. It meant 'improvement for better or worse'.

## WHAT OUGHT TO BE DONE?

The disagreement about whether Britain ought to become more dependent on road transport, or less, will not be quickly or painlessly resolved. Consensuses are not to be had for the asking. Nor can they be imposed by the fiat of central government. They can only be achieved through protracted argument and persuasion.

Despite their deficiencies in current practice, the various

existing procedures of inquiry can with appropriate modification provide excellent forums for the exercise of consensus building. First, the pretence of 'independent' inquiries into questions that are the responsibility of the minister ought to be abandoned in favour of the older, more honest ministerial inquiry. Secondly, and more importantly, the absence of a coherent transport policy should be openly accepted at inquiries; policy conflicts should be confronted not evaded. Inquiring into policy conflicts will not necessarily resolve them, but it stands a better chance than pretending they do not exist.

Central government policy must be settled centrally. But if it is to gain wide support it must be alive to the needs and aspirations of those upon whom it impinges. Much wider discussions of the government's transport policy, summarized in inquiry reports, would greatly help the government to keep abreast of public attitudes. Discussion of both appraisal methods and government policy ought to be positively encouraged at inquiries. Inspectors and committees of inquiry would remain intermediate judges of the issues discussed. The final decision about the matters under discussion would remain the responsibility of central government. Bias would not be removed from the reports of such inquiries but it would be made much more explicit.

The general rule that local decisions should be settled within the context of national policy remains valid. But where a coherent national policy does not exist, a converse rule should apply. To discuss policy in the context of specific transport projects that would impose hardships on some individuals and confer benefits on others, would bring down to earth the Department of Transport's concern with the abstraction called 'Progress'. Those charged with formulating policy have shown that they are well aware of some of the consequences of further increases in personal motorized mobility. In its 1976 transport policy discussion paper the Department declared: 'As car ownership spreads, schools become larger, hospitals are regionalized, out-of-town shopping centres multiply, and council offices are situated farther away; meanwhile the local shop and most offices have often disappeared.'

They are able to estimate the magnitude of some of these consequences statistically. But what these consequences mean to the people affected, they can only begin to understand by listening to these people discuss and argue about them in plain English.

## REFERENCES AND NOTES

1  Department of Transport and Department of the Environment (1978), 'Report on the Review of Highway Inquiry Procedures', HMSO, Cmnd 7133.
2  Department of Transport (1978), 'Report of the Independent Committee of Inquiry into Road Haulage Operators' Licensing' (chairman, Prof. Christopher Foster), HMSO.

3 Copies of the full text of the 'Peeler Memorandum' are available from Transport 2000, 40 James St, London W1.
4 Goldstein, A. (1975), 'Highways and Community Response', Rees Jeffreys Lecture to the Royal Town Planning Institute.
5 Miss Mary Burr.
6 Thompson, E. P. (1978), The State Versus Its 'Enemies', 'New Society', 19 October.
7 The case and its background are reported in the 'Hampstead and Highgate Express', 19 August 1977 and 27 January 1978, 'The Guardian', 17 January 1979, and the 'Evening News', 16 January 1979.

# Part III

# THE VISION

# 13 TAKING THE WAITING OUT OF WANTING

The dominant objective of a transport facilities plan . . . is to reduce travel frictions by the construction of new facilities so that people and vehicles . . . can move about within the area as rapidly as possible. 'Chicago Area Transport Study', 1959

This statement exemplifies the postwar transport planning orthodoxy. It is now being challenged by an alternative view:

Access, not movement, is the true aim of transport. One may have access to facilities without moving much at all. An immobile person may have water and gas at the turn of a tap and electricity at the flick of a switch, have his refuse collected, receive calls from his doctor and deliveries from the shops, talk to his friends on the telephone, all without stirring from his house. . . . The act of travel, with the time, cost and personal effort involved, is something which he usually would prefer to avoid. The true goal of planning, the real meaning of mobility, is therefore access.

This argument is advanced in 'Changing Directions',(1) a report highly critical of the government's transport policies. Similar sentiments are to be found in the government's recent transport policy White Paper, suggesting that it is now well on the way to becoming the new orthodoxy.

The difference between the old orthodoxy and the new is about means, not ends. Contrary to the charge that is sometimes made, the old school did not really believe in whizzing about aimlessly just for the sake of it. If they failed on occasion to state that they were facilitating mobility in order to facilitate access, it was because they thought it was too obvious to need saying. The central criticism of the emergent new conventional wisdom is that transport policies of the 1950s and 1960s pursued the objective of access ineptly. In their obsession with promoting access by road they were heedless to such things as the costs of accidents, pollution, and congestion, the loss of access to local shops unable to withstand the competition of hypermarkets, the social divisiveness created by the loss of access by those dependent on public transport, and the profligate consumption of land by the motorized society that was being created. But there is no disagreement between the old orthodoxy and the new that the true goal of planning is access.

More than sixty years ago E.M. Forster wrote a short story about this goal called 'The Machine Stops'. It was a story about a world in which mechanical progress has run its course. This is how it begins:

Imagine, if you can, a small room, hexagonal in shape, like
the cell of a bee. It is lighted neither by window nor by lamp,
yet it is filled with a soft radiance. There are no apertures for
ventilation, yet the air is fresh. . . . An armchair is in the
centre . . . there sits a swaddled lump of flesh - a woman,
about five feet high, with a face as white as fungus. It is to her
that the little room belongs.

   There were buttons and switches everywhere - buttons to call
for food, for music, for clothing. There was the button that
produced literature. . . . And there were of course the buttons
by which she communicated with her friends. The room, though
it contained nothing, was in touch with all that she cared for
in the world.

In the story everyone's cell was identical, and people rarely left
them:

Few travelled in those days, for, thanks to the advance of
science, the earth was exactly alike all over. Rapid intercourse
from which the previous civilization had hoped so much, had
ended by defeating itself. What was the good of going to Pekin
when it was just like Shrewsbury? And why return to
Shrewsbury when it would be just like Pekin?

The Machine provided access - direct, unlimited access to mankind's desired ultimate ends. It thereby rendered redundant the
necessity for access to the multitude of intermediate ends with
which our civilization is so preoccupied. Access to shops to
obtain food and clothing, access to training to acquire employable
skills, access to recreational facilities to obtain a respite from
work, and access to work to obtain the money with which to purchase access to these things - all such concerns had lost any significance. Mechanization had set humanity free, within mortal
limits, to devote itself exclusively to its ultimate ends.

   The result was a civilization of intellectuals in pursuit of abstraction. And despite its facilities for instant communication and grati-
fication of all material wants, it was always irritably pressed for
time, the almost infinite disproportion between what was accessible
and what it was possible to digest either physically or mentally,
created an endemic frustration that could not be appeased. There
was also a pervasive, though rarely articulated, anxiety about
the purpose of it all:

No one confessed the machine was out of hand. Year by year it
was served with increased efficiency and decreased intelligence.
The better a man knew his own duties upon it, the less he
understood the duties of his neighbour, and in all the world
there was not one who understood the monster as a whole. Those
master brains had perished. They had left full instructions it is
true, and their successors had each of them mastered a portion
of those directions.

It is a bleak and dreary tale. Judging by the neglect in which it
has been left lying it is not reckoned in critical circles to be
great literature. It manages no tension or excitement. The hero's
attempt to escape is obviously futile. Its characters are made of

dull grey cardboard. It grinds its doom-laden way to its predestined end. The medium is the message - mechanical progress is soul-destroying. It is what the critical circles might call a dismal read.

Nevertheless, 'The Machine Stops' deserves a revival. It anticipates in a remarkable way the methods and aspirations of contemporary social engineering. The following is taken from an article in 'Time', 20 February 1978, enthusing about the way in which the silicon chip will enrich all our lives. Despite its remarkable similarity to the world of 'The Machine Stops' it acknowledges no debt to Forster:

It is 7:30 a.m. As the alarm clock burrs, the bedroom curtains swing silently apart, the Venetian blinds snap up and the thermostat boosts the heat to a cozy 70°. The percolator in the kitchen starts burbling; the back door opens to let out the dog. The TV set blinks on with the day's first newscast: not your Today show humph-humph, but a selective rundown (ordered up the night before) of all the latest worldwide events affecting the economy - legislative, political, monetary. After the news on TV comes the morning mail, from correspondents who have dictated their messages into the computer network. . . . Barring headaches, tummy aches and heartaches, the American day should proceed as smoothly as it begins.

It was during the 1960s that the scientific planning of the transport and communications systems of visions such as this emerged as an academic discipline called 'regional science'. Regional scientists established their own association, their own journal, and held their own conferences. Other disciplines, notably geography and economics, resisted this invasion of their territory and countered with works of their own under such headings as locational analysis, land economics, transport and land-use modelling, and urban system theory. Together with the regional scientists they produced and refined the theories and computer models that today constitute the directions for the development and operation of the transport and communications structures through which we interact with one another.

'Methods of Regional Analysis',(2) first published in 1960, is the original basic textbook of access planning. Its approach, it asserts:

may be said to resemble an approach frequently used by physical scientists. For example, Boyle's classic studies of the effects of pressure and temperature on the volume of gases were essentially investigations into the behaviour of masses of molecules; the movement of any individual molecule was not a matter of inquiry.

The single most significant achievement of this approach is the development of a mathematical model of interaction known, from its similarity to the original equation of Newton, as the gravity model. The force of attraction exerted by one body on another was, according to Newton, directly proportional to the product of the masses of the two bodies and inversely proportional to the distance separating them. So also with masses of people. The planner's gravity model assumes that the amount of interaction between two masses of people will be directly proportional to the force of attraction that they mutually exert. By means of this assumption one arrives at a mathematical formulation for explaining the interaction ($I_{ij}$) between two masses of people that looks almost exactly like Newton's:

$$I_{ij} = G\frac{P_i P_j}{d_{ij}^b}$$

Where $P_i$ and $P_j$ are measures of the population masses of two places i and j

   $d_{ij}$ is a measure of the distance between them
   b is a constant that describes the rate at which interaction decreases with distance, and
   G is a constant that translates units of distance and units of population mass into units of interaction.

This model, complicated by a great many operational refinements but with relatively minor modifications of principle, has seen service almost everywhere in the world where access planners and computers have come together. It was used by the Roskill Commission in its endeavours to find a site for a new London Airport, and by the Greater London Council in the planning of its urban motorways. It has been employed in Atlanta and Arras, in Brighton and Buffalo, in Kansas City and Kuwait: it has become *the* model of urban transport planning. Its use has also spread to a wide number of other more specialized applications, such as the planning of shopping centres and hypermarkets and the design of traffic management schemes. The most ambitious use to be made of it so far in Britain is not yet complete. The Department of Transport is proceeding with the development of a £6 million version of the model for use in planning future traffic flows along the nation's major transport arteries.

   Travel, in the language of the access planners, is a disutility, an unwanted but necessary means to an end. It has a cost consisting of time effort and, commonly, money. Since most of the molecules (i.e. people) represented by $P_i$ and $P_j$ craving access to one another do not have unlimited quantities of these to spend, these costs impose a constraint on their interaction. It is this constraint that makes the patterns of mass interaction to some degree predictable. But because this constraint consists of disutilities it is something that, by definition, the planner strives unceasingly to minimize or, ideally, to remove altogether.

   Very few stones are left unturned in the search for ways to reduce the disutility of travel. 'The Need for Route Guidance' a report published by the Transport and Road Research Laboratory,(3) reached the conclusion that car drivers frequently failed to take the shortest route when trying to get from A to B. As a consequence, it estimated, there are 4 per cent more route kilometers travelled every year than is necessary, and if this wastage could be eliminated the saving to the nation would be worth about £470 million per year (net of fuel tax) at 1975 prices. The report conceded that some of this 'wastage', was deliberate, caused by idiosyncratic preference for routes that were inefficient. But most of it, the report argued, could and should be eliminated by a national system of automatic route guidance. The main features of this solution to the problem, estimated to cost over £600 million, are described as follows:

Each participating vehicle would be equipped with a small display unit placed high on the dashboard. At the start of a journey the driver would look up the code for his destination in a book not unlike a telephone directory and would dial it into the display unit, which would store it for the remainder of the journey. The driver would then set off in approximately the right direction. When he approached the first equipped junction the display unit would transmit his destination code to a roadside equipment [sic]. The roadside would reply with codes which cause the vehicle display unit to show a simplified plan of the approaching junction, with the recommended route prominently superimposed. Thus he would be passed on from junction to junction on as near optimum a path as the density of equipment junctions permits.

Automatic route guidance would be but a small step towards the achievement of the ultimate goal of an 'isotropic communications and information surface over the entire earth' envisaged by the proponents of Ecumenopolis cited in Chapter 3. But it is indicative of the sort of computerized direction that would be required for its achievement. Another indication is the system of traffic control employed by London. The flow of traffic in a wide area of central London is now governed by computer-linked traffic lights and monitored by means of television cameras at all major intersections. These television cameras of course permit the police to monitor everything that appears on the screens of the batteries of television sets in the central control room in Victoria.

The disutility of travel and communications can never be reduced to zero; the equipment that is required for modern systems of transport and telecommunications is vastly expensive. But what can be reduced almost to zero by this equipment, in theory at least, is the difference in the time cost of travel between pairs of points on the earth's surface. This has already been achieved for telecommunications. Concorde and developments such as plane-trains and ballistic transports discussed in Chapter 1 permit extremely high speeds over long distances, while speeds between points close together are constrained by road or terminal congestion, and by the rate at which human beings can safely be accelerated and decelerated.

It is frequently contended by transport and communications futurologists that congestion will be greatly reduced in the future by the substitution of telecommunication for much physical travel.

So far this has not happened. On the contrary the growth of
telecommunications traffic has promoted an increase in physical
travel. As the graphs of Chapter 1 show, the growth of telecommunications traffic has everywhere been accompanied by an increase, not a decrease, in physical travel. Most electronic travel
is not a substitute for physical travel but a complement to it,
and frequently for the purpose of facilitating it. In 'The Machine
Stops' Forster forecast that ultimately the growth in physical
mobility would be self defeating because it would obliterate most
of the differences in the world that constitute the principal motive
for travel. Only to the degree that this happens is electronic
travel likely to become a substitute for, rather than a complement
to, physical travel.

Although the ideal of the isotropic plane has yet to be achieved,
the directions for coping with the interactive consequences of
such an achievement are already available. If distance becomes
of no significance, we simply strip it from our gravity model.
(This can be done by allowing b, the exponent of distance, to
shrink to zero, thereby leaving the model with a denominator of
1.) We are left with a model in which the amount of interaction
between two places is determined by their population masses -
and by nothing else.(4)

The molecules of which these masses are comprised are assumed
to be, like the citizens of Forster's Machine, 'seraphically free
from taint of personality'. 'Methods of Regional Analysis' puts it
in rather dustier language: 'let there be no significant differences among subareas in the tastes, incomes, age distributions,
occupational structures, etc., of their populations'. Where such
differences interfere with the smooth running of the Machine, we
are told in 'Mental Maps',(5) another example of the genre, the aim
should be 'to smooth out the hills and valleys of the perception
surfaces to make them perceptual plateaus, or flat administrative
chessboards upon which people can be assigned'.

Theoretical advances since 'Methods of Regional Analysis' provide additional assitance in imagining the end state towards which
access planning aspires. In 'Entropy in Urban and Regional Modelling'(6) it is demonstrated that the gravity model can be mathematically related to the Second Law of Thermodynamics. Interaction
between masses of people is, it seems, theoretically virtually
indistinguishable from the entropy maximizing behaviour of masses of gases, and is capable of prediction using similar models.

When distance has been abolished, geographical distinctions
obliterated and cultures made uniform, social entropy will be
maximized. Patterns of interaction will be random. Freedom of
access will be total. Dependence on the machine that produces
this freedom will be absolute. The world will be populated with
mechanics, and the Machine will be out of hand. This is how
Forster imagined it: 'Science retreated into the ground to concen-
trate herself upon the problems that she was certain of solving.
. . . Men seldom moved their bodies; all unrest was concentrated
in the soul.'

We have, of course, no intention of actually getting to such a state. We all, doubtless, agree with Robert Louis Stevenson: 'To travel hopefully is a better thing than to arrive.' The trouble is, if we travel hopefully for long enough in the wrong direction, we are bound to end up where we do not want to be.

REFERENCES AND NOTES

1   The Independent Commission on Transport (1974), 'Changing Directions', Coronet Books, London.
2   Isard, W. (1960), 'Methods of Regional Analysis: an Introduction to Regional Science', MIT Press, Cambridge, Mass.
3   Armstrong, B.D. (1977), 'The Need for Route Guidance', Transport and Road Research Laboratory, SR330.
4   Such a development can also be described in terms of the centred interaction fields discussed in Chapter 6. The centred interaction field model is, in effect, a form of gravity model from which the effect of population mass has been removed. The net effect of reducing the difference in the cost of travel between all pairs of points in the world to zero would be to transform the bell-shaped dome of the model to a perfectly flat pancake.
5   Gould, P., and White, R. (1974), 'Mental Maps', Penguin.
6   Wilson, A.G. (1970), 'Entropy in Urban and Regional Modelling', Pion.

# 14  THE NATIONAL HEALTH*

As planners we provide the framework within which people choose how to develop the most satisfactory lives for themselves.(1)

This statement, taken from an editorial in 'Environment and Planning', is a frank and accurate description of the way most planners view their job. Abler et al.(2) take a similar view of planning and suggest that the role of social framework provider is analogous to that of doctor:

In the same way that we want the health professions to manipulate our experience to prevent us from becoming ill or to cure us if we do, we ask other scientists to perform similar manipulative functions. We require economists to help us understand and adjust our corporate and national economic systems. We wish to manipulate economic activity to keep conditions compatible with chosen criteria of economic health. Aeronautical engineers are expected to provide us with more rapid means of travel. . . . No matter what branch of science we consider, the demand for diagnostic, prescriptive and preventative activity is similar. We want scientists to prevent us from experiencing unpleasant events and to structure the future in such a way that we will experience pleasant events in as great an abundance as possible.

The doctor analogy is apt. Planners do see themselves as distinguishable from the man in the street. They see themselves, and are seen by the general public, as doctors, as people with specialist knowledge not available to, and often not comprehensible to, the lay public. They see themselves, and would like to be seen by others, as repositories of knowledge and skills to whom an ailing society turns for help.

The relationship between planning doctor and patient can be illustrated by the investigation of the Roskill Commission(3) into the congestion of London's air transport arteries. The Commission, which conducted one of the largest and most comprehensive transport inquiries ever undertaken anywhere in the world, clearly saw its role as that of social doctor and was greatly concerned that its prescription concerning a new airport should accord with, the spirit of the Hippocratic oath. Their concern, the commissioners insisted, was with 'the overall national interest', and they observed, '. . . if our final recommendation were to command

*A slightly modified version of a paper first published in 1977 in 'Environment and Planning A'.

respect and acceptance it had not only to be as right as the best methods could make it but the reasons leading to our judgement had to be as objective and as explicit as we could make them'. But like a doctor treating a lay patient the Commission remained the judge of what evidence was relevant to its diagnosis. 'We decided on the short list (of airport sites) on the best evidence then available. Public participation could thus be concentrated on a short list of sites and the crucial issues debated with reference to them alone.' Inviting the patient to participate in the preliminary diagnosis of his ailment, they believed, would have prolonged their proceedings unduly.

Like a doctor, the Commission obtained some of its evidence by observing the patient, and some by asking him to describe his symptoms. The Commission, also like a doctor, tended to have a rather low opinion of the evidence obtained by the second method. When, for example, the Commission tried to find out what cash value people placed on the 'community life, friendships, and other local associations' that would have been disrupted by a new airport, they were particularly suspicious of people who told them that it was incalculable. The general public, like a hypochondriac, is frequently though to have ulterior motives when presenting symptoms to his doctor.

## THE TRANSPORT PLANNER'S CRITERION OF HEALTH

The general public can be a patient who is very hard to please. If decisions are to be taken to promote its over-all best interest, then criteria for deciding what this interest is must be defined. In their description of the duties of the aeronautical engineer, quoted above, Abler et al. presumably chose an illustration that they considered uncontentious. That the speed of a society's means of travel would serve as a measure of its progress appears to them obvious.

This is an example of what might be termed 'the greatest mobility for the greatest number criterion'. It is found in a number of different guises and is the criterion that dominates, to the exclusion of almost all others, the work of transport planners. It guided the work of the Roskill Commission, who argued that 'this country should not purchase peace and quiet at the price of cutting itself off from the world's air routes': it informs the planning activities of the British Airports Authority, the Civil Aviation Authority, and that part of the Department of Trade responsible for airport planning. It is vigorously defended at motorway inquiries by the Department of Transport, and lip-service is paid to it by British Rail and other public authorities presiding over the nation's rapidly dwindling public transport services. As noted in Chapter 12, it is also found in transport studies that express concern about present trends in mobility; the most common charge levelled at the car and airplane is that they are inegalitarian modes of transport and impede progress towards the goal of greater mobility for all.

## ABSURD SCENARIOS

Although many transport planners could no doubt be found who would reject Sir Peter Masefield's ballistic missile scenario as absurd (p. 4), their reasons for doing so would be interesting to know. For so long as they seek to maximize transport efficiency and employ mathematical models in order to minimize transport costs, albeit subject to environmental constraints, their criterion of health is essentially the same as Masefield's. If only the technology were available to whisk the frustrated traveller on his way - comfortably, noiselessly, fumelessly, cheaply, efficiently, and unobtrusively, that is, by some very successful variant of high-speed mass transit - then the speed of light would appear to be the limit to the progress towards which they all aspire. They would appear to differ only in the degree of their optimism that technology can successfully phase out the undesirable side-effects.

The principle that efficiency generates traffic is very well established. The trends in mobility described in Chapters 1, 2 and 6 can only be explained by rising affluence, and by developments in transport technology that have greatly decreased, both relatively and absolutely, the costs of travel.(4) If the transport planner aspires to more efficient, lower-cost transport, he aspires to more traffic. He cannot be in favour of one and against the other.

The Roskill Commission was unashamedly in favour of both. If new airport capacity were not built, they argued, Britain would risk losing tourists to other countries, economic growth would be inhibited, flying would remain forever the prerogative of a well-to-do minority, and, perhaps most important of all, the nation's pre-eminence in the field of international aviation would be threatened. About this last threat the commissioners felt strongly. 'The phrase "London is the Clapham Junction of the air" was no mere chauvinistic cry. It represented the deeply and sincerely held views of the economic and political importance of maintaining the position of this country as one of the two foremost aviation nations in the western world and as the leading aviation nation in Western Europe'.(5) Any inefficiencies that threatened to reduce the rate of growth of something so desirable would clearly be unhealthy.

The absurdity of the greatest mobility criterion lies in its lack of any limits. Sir Colin Buchanan, quoted in Chapter 1, now has the whole of Europe at his disposal for his holidays and thinks that it should be at the disposal of all classes. With Sir Peter Masefield's anywhere to anywhere ballistic transport, the whole world would be at his disposal and it would clearly be unjust for all classes not to be invited along. The vision of a future in which the whole world is accessible to the whole world is contemplated with apparent satisfaction by the makers of Concorde, the 'earth shrinkers' of British Airways, and the planners of Britain's airports.

Where might it end?

Transport planners seem to find themselves with much the same dilemma as economists. The graph of increasing mobility has served as their measure of progress in the same way that the graph of rising gross national product has served the economist. The two measures are intimately related. Because the correlation between the indices is so close, forecasts of GNP have become the dominant ingredient of the air traffic forecasting model of the Department of Trade(6) and the freight forecasting model of the Department of the Environment(7), GNP subsumes the mobility indices in the sense that transport activities are a constituent of the grand sum of national economic activity. But the relationship is even closer. Because personal travel is essential to the conduct of most economic activity, and freight traffic is the link between production and consumption, mobility is an essential aspect of GNP.

The close connection between these two conventional measures of progress permits us to visualize in some graphic detail the state towards which we are progressing. In recent guidance to its road planners the Department of the Environment(8) indicated that it is the government's policy that progress, as measured by an increasing GNP, should continue in 'real terms' at 3 per cent per year 'in perpetuity'. Since perpetuity is such a very long way off, the scenario that follows is based on the rather arbitrary chosen year of 2205.

The year 2205 is a milestone in so far as it is the year in which Britain becomes a millionaire society. It is the year in which, assuming that the government's growth target is achieved, average incomes will reach 1 million pounds. The scenario also assumes that a socialist government will have been in power for most of the intervening 231 years, so that deviations about the average income will be extremely small. For all practical purposes then, in 2205 everyone will be an income millionaire. We can, of course, only hazard a guess what life will be like. Clearly, because millionaires are not inclined to do menial work for others, most of the services that can now be bought by millionaires will have disappeared from the market; other goods, such as peace and quiet, that people could not afford when they were poorer will also not be for sale at the old prices.

We can, however, using the Department of the Environment's freight forecasting model, fill in a few important details. The volume of freight moving about on the roads will have increased one-hundredfold. To accommodate this our descendants would need 60 million lorries.(9) This assumes that the average carrying capacity of a lorry remains no larger than that of today's lorries, and this means that the number of lorries would almost exactly equal the population. Since only about half the population can be expected, even then, to be qualified drivers, we might, as a tentative adjustment to our scenario, double the lorry-carrying capacity and halve the number of lorries to 30 million. This would make the average lorry about the size of one of today's juggernauts. A picture of daily life in the millionaire society begins to emerge. Since the service sector of the economy will have virtually disappeared people will spend most of their time driving around in the family juggernaut picking up piles of machine-made stuff from automatic warehouses or wandering about the tarmac plain searching in vain for someone to carry out repairs when it goes wrong. Indeed, such is the volume of stuff that will require shifting that it is doubtful whether they will have the time to do all the holiday to-ing and fro-ing expected of them by the road and airport planners.

## A DEFECT

As a criterion of health the greatest mobility standard has a defect that cannot be 'phased out'. It is a defect that becomes more obvious as mass mobility increases: increased efficiency generates traffic in a highly constrained way. Because people are strictly limited in the amount of time they have to spend each day, if they spread themselves more widely they must spread themselves more thinly. If they go farther to work, farther to school, farther to shop, farther to their holiday resorts, and farther to visit friends, then they must be spending less time closer to home. As a consequence, a large number of institutions have grown to keep pace. People increasingly work in larger offices and factories, 'learn' in larger schools, shop in larger stores, and spend their holidays in larger packages. And if they have a larger number of more distant friends they have fewer closer to home.

People also have their lives planned for them on a larger scale than ever before. The reorganization of local government and the growth of the Common Market's bureaucracy are but two examples of planning institutions struggling to keep up with the activities of an increasingly mobile population. If people spread themselves more widely and the planning institutions do not, then the latter become impotent.

The Roskill Commission was impressed by the size of its task. They described it as 'an investigation into the largest and most important piece of transport investment that this country . . . has ever seen'. And it was a problem that, if measured by the volume of air traffic to be served, they expected to double in size every seven or eight years. It was a task that appears to have caused the commissioners quite genuine distress. They were confronted at every turn by 'dilemmas'. They thought Brent geese were desirable and they thought economic growth was desirable. They thought that both Stewkley church and an air traveller's time were worth saving. They valued the peace and quiet of people living not only near Cublington, but near Thurleigh, Nuthampstead, and Foulness as well. They abhorred the destruction of communities, the loss of cultural heritage, and even the probable loss of life that would result from building a new airport. But even more they abhorred disappointing the British Airport Authority's 'silent hosts of future passengers'.

Their task demanded more concern than it was humanly possible to give. They were confronted with not just one choice to try their compassion but a bewildering multitude of choices. The lives of far more people than they could ever hope to know in an intimate way would be affected by their decision. The pleasures and sufferings of these people were in their hands, but at best they could hope to know them only in a distant statistical sense. Their compassion, like the activities of the people being planned, had to be spread widely, and thinly.

The commissioners' answer to their dilemmas was cost-benefit analysis. They idenitified all the factors that they felt were relevant to their decision, measured them in whatever units seemed most appropriate, and then transformed them into cash. The measurement of the selected factors, conducted with the assitance of an impressive array of mathematical models, produced, for each of the sites being considered, a large collection of imcompatible numbers; decibels, minutes, houses, flights, schools, acres, private airfields, hospitals, miles, and deaths are just some of the units in which they measured their factors.

With the help of even more impressive models all of these were then transformed into money, because money, according to the Commission, 'is clearly the most convenient the most readily acceptable, and the most easily understood general standard of comparison'. To do otherwise, they argued, would have been to substitute implicit intuition and prejudice for explicit valuation and objective analysis.

The Commission's methods were heavily criticized and their recommendation conspicuously failed to command the respect and acceptance that they had hoped for, but the commissioners remained unrepentant. 'Nothing has happened', they said, 'to make us regret our decision [to use cost-benefit analysis]'. Most of the damaging criticisms levelled at their method they did not attempt to answer. They simply asserted in aggrieved tones, 'No one has yet suggested a better alternative'.


DECISION-MAKING

The problem of making planning decisions troubled the late Richard Crossman(10) and he discussed his alternative in an illuminating way in his diaries. He recounted an argument that he had with his civil servants who tried to dissuade him from going to see 'the actual situation' before making up his mind. To do so for any one case, they had argued, would be to establish a precedent for all other cases and might expose him to new evidence that had not been available to those writing the reports upon which he was expected to base his decision. The danger that his civil servants feared was not that he might improve his understanding of the question at issue but that he would inject an element of inconsistency into the decision-making process. Because he could not get personally involved in every decision required of him it

was important to ration his involvement equitably, and this meant
dealing only with those 'facts' that had filtered through the civil
service system to him. That his civil servants' fears were not
without foundation is illustrated by this passage:

> Before I became a Minister I used to worry, wake up early in
> the morning in a panic about whether I had done something
> wrong. Now I have so many more things to worry about, so
> many big decisions to take, I find myself worrying much less.
> . . . I find the job of a Minister relatively easy. When I sit
> at my desk at Prescote and pull out a mass of paper from the
> red box and see that I have to decide on the boundaries of
> Coventry or on where to let Birmingham have its new housing
> land, I find these decisions easy, pleasant, and take them in a
> fairly lighthearted way.

Alan Wilson, in an article discussing computerized approaches
to decision-making, observes that the growth of computer-based
planning has been accompanied, particularly in the realm of transport planning, by a worsening of many of the problems the
planners have been attempting to solve. But he warns that the
methods are not to blame:

> During the 1960s, transport engineers trying to counter traffic
> congestion started to use computer models. They seemed a
> planners' dream. It became possible to make forecasts, to test
> alternatives and to select what was seen as a best plan. Mean-
> while, for many consumers, traffic is becoming a nightmare.
> They are aware of increased congestion, declining public trans-
> port services, traffic management schemes diverting more traf-
> fic past their front doors and motorways ruining their back
> gardens. Is this all a consequence of planning by computer or
> a coincidence? I maintain it is a coincidence, and that models
> are essential tools for the planner and can be in the consumers'
> interest.(11)

The failures of transport planning he says are due not to using
quantified models in the analysis, but to weaknesses in the original policy-making, in design, and in communication between politicians, planners, and the public. The only alternatives to computer models for dealing with complex problems, he argues, are
'bigotry' and 'guesswork'. Doubtless he would cite the Crossman
method for making decisions as a combination of the two.

But the intuition and prejudice that the Roskill Commission
shunned, and the guesswork and bigotry that Wilson warns us
against, cannot be got rid of so easily. Computer models do not
get rid of conjecture and unreasoning attachment to creed, rather,
they systematize, it. They systematize it in a peculiarly bigoted
way. They require that the planner's criteria of health, presumably to be supplied by the politician, be operationally specifiable,
that is reducible to numbers; and they are blindly intolerant of
all considerations that cannot be neatly specified in this way. The
satisfaction in the lives led within the planner-provided framework must be measurable, or the planner's computer is of no use
to him. This presumably accounts for Wilson's description of the

inhabitants of this framework as 'consumers'. The framework in
this computerized scheme of things becomes analogous to a giant
chain of warehouses in which the 'quality of life' is measured by
the range of consumer choice and total annual sales.

But, it might be argued, this presents a misleadingly materialistic view of things. Values and the criteria of health might
change. In the future new 'aesthetic goods' such as peace and
quiet, beauty, clean air, and community life might appear on the
shelves in great profusion and make an increasingly important
contribution to the grand total of annual sales.

Perhaps the proportion of GNP made up by the production and
consumption of material goods might decline while GNP continues
to rise. If it does the measurement of progress will become an
increasingly uncertain exercise. For example, should the increasing cost of services and aesthetic goods be interpreted as an upward shift in the value that people place upon them, or simply as
an inflationary development requiring a numerical adjustment to
the measuring rod of progress?

Such questions constitute the research frontier of quantitative
planning. As shown in Chapter 11, they are not academic questions of concern only for more affluent generations yet to come.
They are with us now, and they have already received some tentative answers. The Roskill Commission decided that the 'real' value
of people's time and 'recreational consumer surplus' would increase
in step with 'real incomes' at 3 per cent per year. The real value
of peace and quiet, they thought, would increase rather more
quickly; its growth rate they put at 5 per cent. In the cost-
benefit studies of road projects undertaken by the Department of
the Environment, the value of a life (£14,960 at 1968 prices) is
assumed to increase at 3 per cent per year, again in 'real terms'.

As roads, airports, and economically productive activities encroach on less productive parts of the country the price of beauty
and tranquility will rise and they will become the exclusive possessions of those who are willing and able to pay the going price.
It is not until people become rich enough to buy aesthetic goods
that they become of any economic importance. If someone has
acquired them free as part of his birthright, even if he declares
them to be his most treasured possessions, if he is not willing and
able to spend money to defend them, a cost-benefit decision-
making model can acknowledge their existence only at the threat
of its own destruction.

The prices used by cost-benefit analysts have a clear and
direct relationship to the mobility landscape depicted by Figure
6.5 on p. 92. The landscape depicted by the mobility curves of
Figure 6.5 is one that in its major features correlates highly with
a number of other statistical landscapes. People who are wealthy
can afford to buy, and do buy, more mobility than people who are
poor. They can also afford to pay for safe and pleasant environments in which to live, and this is reflected in the price of their
property. For example, inflicting a new burden of traffic noise

on a neighbourhood will have a dramatic effect on property values
if it is a very wealthy neighbourhood, and a negligible effect if
it is a very poor one. Thus one could draw maps showing spatial
variations in socio-economic variables such as income, property
values, sensitivity to noise, the value placed on time, or even the
cash value placed on a human life. If one did, one would almost
certainly find that the patterns on these maps corresponded
closely to the pattern of the mobility landscape.

The implications for planning based upon these values are clear.
Other things being equal the transport projects that flatten the
flattest mobility curves (i.e. projects serving the transport desires
of the most wealthy) will yield the greatest benefit. And building
motorways through, or flying airplanes over, the zones of lowest
mobility (i.e. the poorest neighbourhoods) will minimize the costs
of these projects. In brief, a consistent application of the optimizing principles of cost-benefit analysis would accentuate already
existing disparities.

Computerized, cash-quantified decisions are not objective and
unprejudiced. They are arbitrary and bigoted. They depend upon
a linear index of progress. That index, the measuring rod of
money, has become a stick with which we are prodding ourselves
down the road towards the tarmac plain of 2205.

## ALTERNATIVES

Both the Crossman method of making decision and the cash-
quantified alternative involve bigotry and guesswork. The Roskill-
type method has the advantages that it is more consistent and that
its criteria of health are more explicit. But, because its criteria
are economic, the decisions it yields will lead inexorably in the
direction of the millionaire society. The lighthearted Crossman
method, which is the traditional method, has the advantage that
it does not preclude the possibility of taking 'aesthetic disbenefits' into consideration, but this advantage is gained at the cost
of consistency. Both methods, applied to large planning problems
are bound to strike most of the people upon whom the decision
impinges as arbitrary and unjust. The reason for this was ack-
nowkledged by the Roskill Commission.

The commissioners recognized that public participation in the
decision was important if their recommendation were not to seem
an arbitrary imposition. But their attempts to foster a sense of
participation failed. Observing experts arguing about esoteric
models at the public hearings did not fill many people with a sense
of participation. Neither did most people feel that having a cash
value attached to their 'householder's surplus' or their favourite
fishing stream was a very effective way of participating in the
decision. If the decision were to command the acceptance and
respect that the Commission sought, the people upon whom it was
to be imposed had to feel that it was their decision.

The public inquiry into a stretch of the proposed M25(12) motor-

way near Epping in Essex provided a demonstration of the impossibility of genuine public participation in decisions of this
scale. The inquiry opened, in a small but packed public hall in
Epping, with a heated and at times tumultuous argument about
whether or not the Department of the Environment had fulfilled
its obligation to give adequate publicity to the general effects of
its proposal. A man in the audience, impatient with the semantic
squabble to which the lawyers were reducing the argument,
shouted, 'Of course they haven't published the "general effects"
of their proposal. If they had they would have had to hold their
inquiry in Wembley Stadium.'

This put the problem in a nutshell. The capacity of Wembley
Stadium is a gross underestimate of the numbers of people who
are affected in one way or another by modern, large-scale transport projects. The problem of merely informing people of the way
in which their interests will be affected, let alone of reconciling
the multitudes of conflicts of interest involved, is insurmountable.
The decision, to most of those affected, must seem arbitrary.

This is a problem common to almost all aspects of planning.
Regional structure planners also face an impossible task in involving 'the public' in the formulation of their plans. For most people
the development of a sense of involvement in planning requires
their knowing what the various proposals being considered will
mean to them personally. But this is specifically precluded by
the procedural rules that require the discussion of plans in the
abstract. A Structure Plan, the Department of the Environment(13)
has instructed, must not descend to a detailed description of the
way in which individual members of the general public might be
affected: 'In dealing with key issues, precise areas should not
of course be defined.' Thus it is not the general public who will
participate in structure planning but a small, self-selected group
with either the cast of mind required to bring the modellers'
abstractions down to earth, or the peculiar mentality of those who
live permanently on the Astral Plane of abstraction and never
come down to earth.

No matter how beautifully gilded the framework might be, if
people do not have a sense of participating in its design and construction they will resent it as a cage. If most people cannot be
consulted about most of the issues that impinge on their lives they
will feel, and be, irresponsible. This has been recognized as a
problem ever since Adam and Eve stole the apple from the Tree of
Knowledge - the one thing lacking in the Garden of Eden was
consumer participation in planning - but it seems to be getting
worse. Planning problems grow ever larger in scale, but it is impossible to consult more people about more issues. Resentment
of planners, and irresponsible behaviour both seem to be increasing. What then is the alternative?

An elaboration of the doctor analogy with which we began provides some suggestions about the direction in which we might look.

Doctors dispense psychotropic drugs; modern planners dispense
models. Both are employed in conscious attempts to modify the

perception of reality, and both are commonly used as a response
to stressful situations. The drug addict cannot cope with his
problems in their raw state, and society is also confronted by
worrying and intractable problems - traffic chaos, crime, pollution, energy crises, and nuclear holocaust threaten to overwhelm
us.

In both cases the drug resorted to produces a comforting sensa-
tion of calm detachment. Problems that in their raw state were
seen as large, complex, and threatening are seen in a more
manageable light. Also both have a pronounced tendency to
require an escalating dosage to achieve the same effect. The Roskill models, for example, gave some people the illusion of coping
with the airport problem, and in so doing threatened to contribute
very directly to the creation of a problem fifteen times larger by
the turn of the century, a problem that would require models
fifteen times more powerful to produce the same soothing relief.

In both cases there is a danger that the dosage will escalate
until a fatal overdose level is reached. The increase in centralized
decision-making assisted by computerized data processing is
symptomatic of an increasing dependence on computer models which
has already reached an alarming level; although they have been in
widespread use for less than ten years they are described by
Wilson as 'essential tools'. Some euphoric modellers envisage the
whole world in a state of global socio-economic integration maintained in a state of homeostasis by a central government using the
ultimate in computer models.(14)(15)

In both cases effective treatment is very difficult. A too sudden
withdrawal of a drug upon which the patient has become dependent produces a 'cold turkey' reaction in which the system concerned experiences a painful upheaval. Also, it seems, there is
no 'technical' cure. There are treatments for drug addicts, but,
until the addict acquires, somehow, the will to kick the habit, he
is almost certain to lapse back into his habit or suffer a nervous
breakdown. Likewise shutting down all of our society's computers
overnight would no doubt produce a catastrophic social collapse,
so great has our dependence already become. Unless, and until,
we can agree collectively to do without the things that only computers can provide we will remain hooked.

We are all junkies, dependent on models in all sorts of ways.
There are a few high priests of the cult of quantified perception
who summon us to seek our salvation in global-scale cybernetic
models. But perhaps most damage is done by the growing number
of pushers who positively encourage society's dependence on the
computerized planning systems they peddle. The true doctor is
distinguished from the pusher by the conviction that too much
quantified perception is unhealthy and produces a distortion of
reality every bit as harmful as that caused by heroin. Reality in
its raw state will no doubt always seem troublesome, but building
a model and sweeping the troubles under its carpet of assumptions
will not make them go away.

The goal of unlimited progress, allied to linear measures of progress such as mobility or GNP, enlarges the framework in which we all live. Although the consequences of global, mass ballistic transport, or of a British GNP grown to 60 million million pounds are unimaginable, progress in this direction requires ever larger systems of planning and control. It requires ever larger and more elaborate models and ever more 'effective' ways of collecting 'inputs' to fill them. It can only intensify the apathy and irresponsibility with which 'the consumer' responds.

Even the most rudimentary of communities must have a framework in the form of an implicit constitution according to which it conducts its affairs. But a sense of community cannot be scientifically planned. The feelings of respect and affection for one's neighbours, and the sense of belonging and of participating in the conduct of community affairs that are essential to a healthy community spirit can, however, be killed by science. Because a sense of community is not something that can be scientifically weighed and measured, alienation, which is the absence of this sense, is not an affliction that can be treated by science. Scientific planning, which deals only in the weighable and measurable aspects of human affairs, has encouraged the growth of institutions beyond the point where a sense of community can exist. In a world afflicted with megainstitutions, the reformed pusher with his quantitative expertise, by suggesting ways in which society might reduce the dosage of mathematical models upon which it has become so critically dependent, can perhaps help; but only if, like a good doctor, he is careful to recognize his own severe limitations and the dangerous potency of the drugs at his command. First, however, the patient must accept that he is ill and find, somehow, the will to be healthy.

## REFERENCES AND NOTES

1   Wilson, A.G. (1972), Editorial, 'Environment and Planning', 4, p. 379.
2   Abler, R., Adams, J., Gould, P. (1972), 'Spatial Organization', Prentice-Hall.
3   Commission on the Third London Airport (1971), 'Report', HMSO.
4   These trends are now threatened by increasing fuel costs. The explanation of most of the 'official' forecasters is that 'normal growth' has been delayed. I am concerned here to discuss not the possibility, or probability, of past trends continuing, but the desirability. The more desirable such trends are held to be, the greater the cost that will be accepted in order to obtain the fuel to sustain them.
5   Such chauvinism may not be 'mere', but whatever sort it is seems to me indistinguishable from the sort that fuels arms races and starts wars.
6   Department of Trade (1974), 'Maplin: Review of Airport Project', HMSO.

7   Tanner, J.C. (1974), 'Forecasts of Vehicles and Traffic in
    Great Britain', Transport and Road Research Laboratory
    Report LR650.
8   Department of the Environment (1975), 'Standard Forecasts
    of Vehicles and Traffic', Technical Memorandum H3/75.
9   The model assumes that if gross domestic product grows at
    3 per cent per year, tonne-kilometres of road freight will
    grow at 2 per cent per year.
10  Crossman, R. (1975), 'Diaries of a Cabinet Minister', Jonathan
    Cape.
11  Wilson, A.G. (1975), Cities, Planners, People and Computers,
    'New Society', 1, May pp. 258-61.
12  Known at the time as the M16. 'A Case Against the M16', pub-
    lished by Friends of the Earth, 9 Poland St, London W1, con-
    tains background to the inquiry, plus a case against the
    proposal.
13  Department of the Environment (1974), 'Structure Plans',
    Circular 98/74.
14  Beer, S. (1971), The Liberty Machine, 'Futures', 3, pp.
    338-48.
15  Meadows, D.H. et al. (1972), 'Limits to Growth', Universe
    Books.

# 15  HOMUNCULUS ECONOMICUS

Lurking in every one of us is an homunculus economicus. He is a beady-eyed little fellow whose job is looking after number one. He is a consumer. If anything is going, he is there to see how much of it he can get for himself. He is extremely well informed. He knows the price of everything and exactly how much of everything he wants at the prevailing prices. He has a sharp mathematical brain and can reorder his wants in a flash if the price of anything changes. He has a voracious appetite that no amount of consuming can diminish. Altruism is incomprehensible to him. He weighs every action in the scales of self-interest and pursues only those that register personal gain. When on occasion his behaviour appears unselfish or co-operative, it will be found on closer inspection to be prompted by far-sighted, 'enlightened' self-interest. But such behaviour is relatively rare. The beady eyes are characteristically myopic, preoccupied with the spotting of bargains close at hand. He is a nasty, egoistical little fellow, and most of us are thoroughly ashamed of him.

He has three common English names: utility maximizer, profit maximizer and economic man; and for one so pre-eminently qualified to look after himself he has acquired a surprising number of volunteer advisers. Economists are altruists with a professional interest in selfishness. They offer their services, generally for a quite modest fee, as detached scientific advisers on optimizing strategies for other people's value judgments. As Nicholas Kaldor pointed out over forty years ago, the efficacy of their advice is crucially dependent on homunculus economicus behaving true to form: 'The scientific status of the economist's prescriptions is unquestionable provided that the basic postulate of economics, that each individual prefers more to less, a greater satisfaction to a lesser one, is granted.'(1)

Although the economist's diagnoses and prescriptions are often of bewildering mathematical complexity, their essence is quite simple. It is summarized by Professor Paul Samuelson in 'Economics', the best-selling economics textbook of all time. Happiness, looked at scientifically, is but a ratio:

$$\text{Happiness} = \frac{\text{Material Consumption}}{\text{Desire}}$$

He notes that Thoreau once recommended people to try to reduce the denominator, but goes on to dismiss this as old-fashioned advice that 'now gives way to insistence on increasing the numerator of material real income'.(2)

Off duty most economists are much more likeable than their patrons, and occasionally embarrassment over the company they keep breaks through the hard professional façade. Keynes tried to clear his conscience in a much-quoted essay written in 1930 entitled Economic Possibilities for our Grandchildren.(3) In this essay he tried to persuade us that homunculus economicus really is hungry, and if only we keep feeding him exponentially increasing amounts until about the year 2030 he will stop pestering us:

> I see us free, therefore, to return to some of the most sure and certain principles of religion and traditional virtue - that avarice is a vice, that the exaction of usury is a misdemeanour, and the love of money detestable, that those walk most truly in the paths of virtue and sane wisdom who take least thought for the morrow. We shall once more value ends above means and prefer the good to the useful. We shall honour those who can teach us how to pick the hour virtuously and well, the delightful people who are capable of taking direct enjoyment in things, the lilies of the field who toil not, neither do they spin.

But although this end was in sight for people with vision as acute as his, the road to it, he warned, would be rough:

> But beware! The time for all this is not yet. For at least another hundred years we must pretend to ourselves and to everyone that fair is foul and foul is fair: for foul is useful and fair is not. Avarice and usury and precaution must remain our gods for a little longer still. For only they can lead us out of the tunnel of economic necessity into daylight.

Keyne's message was remarkably like that of Marx: although the means will be nasty the end will justify them. Although less dogmatic than Marx about the inevitability of the end, Keynes nevertheless displayed great confidence in the power and momentum of compound interest:

> The pace at which we can reach our destination of economic bliss will be governed by four things - our power to control population, our determination to avoid wars and civil dissensions, our willingness to entrust to science the direction of those matters which are properly the concern of science, and the rate of accumulation as fixed by the margin between our production and our consumption; of which the last will easily look after itself, given the first three.

Both Keynes and Marx, and their ideological descendants, are, unfortunately, utterly vague about the means by which all human souls everywhere will be transformed, as the culmination of the process of material accumulation, from avaricious and usurious capitalists, or from hardened proletarian dictators, into 'delightful people'. Lenin, some thirty years after Marx's death, had to report that no progress had been made on his side of the ideological fence towards the solution to this puzzle. Nor did he expect much progress: 'By what stages, by what practical measures humanity will proceed to this higher aim - we do not and cannot know.'(4)

Such embarrassments were but a minor distraction to economists
of Keynes's generation. Economists even then were in the habit
of discounting future costs and benefits at the prevailing interest
rate so the problem of a transformation that lay at least 100 years
in the future did not loom large in their calculations. But now,
fifty years on, the problem is beginning to look a little larger
and is becoming, for some economists, a more central concern.
Although most of the world still leads a mean and precarious
existence there is now a multitude sufficiently numerous and sufficiently affluent to serve as data for a preliminary test of the
Keynesian hypothesis, that if only we feed him enough there will
come a time when the homunculus will become contented, generous,
cultured and delightful. The evidence thus far is mixed, or rather
interpretations of the evidence conflict. Although some people
who live in castles and fly in Concordes are of the opinion that
they are more delightful than people who do not, the interpretation of the majority who do not is that they are not. And since
the chances of everyone enjoying a castle/Concorde life-style
even a 100 years from now seem slight, there is growing uncertainty among the ranks of professional economists about how and
when, and even whether, the Keynesian state of economic bliss
might actually be achieved.


## STRESS AND UNCERTAINTY

In 'Economics' (1970 ed.) Samuelson notes that strange symptoms
of stress are appearing in those countries that have approached
closest to a Keynesian state of bliss; large numbers of people
seem to be engaged in a 'disquieting search for meaning and purpose in life'. This symptom he attributes rather unhelpfully to
'the existential vacuum'. In 'Positive Economics', another popular
economics textbook, Lipsey, discussing what he calls the 'non-
economic costs' of economic growth, acknowledges that 'today
there is debate and uncertainty as to whether or not the final
bill [for economic growth] would include disaster and destruction
for the human race itself'.
    There is no shortage of prominent economists to be found ex-
pressing similar anxieties. There is a burgeoning literature on the
topic. Perhaps the most concise and representative expression of
the current economic angst is to be found in the Dai Dong declaration, 'Towards a Human Economics'. This is a statement sponsored
by twenty-one prominent economists from seven different countries,
which, the sponsors hope, will be signed by thousands of other
economists throughout the world.(5) Their analysis is worth examining for the reason given in the declaration itself: 'The power of
the economists and therewith their responsibility, has been very
great indeed.'
    Their broad view of the problem is set out in the declaration's
first paragraph: 'The evolution of our global household earth is
approaching a crisis on whose resolution man's very survival may

depend, a crisis whose dimensions are indicated by current rates
of pollution expansion, runaway industrial growth, and environmental pollution, with their attendant threats of famine, war, and
biological collapse.'

This state of affairs they point out has been arrived at not
through the operation of inevitable laws of nature but through a
history of decisions in which economists, particularly recently,
have been influential participants. What is needed now, they
argue, is 'a new vision'.

According to the declaration, the principal defect of the old
vision is that 'the economist's traditional measure of national
and social health has been growth'. This vision, permeated by a
spirit of confident purposiveness, can be found elaborated in
libraries full of economic texts and journals. The old economists
knew where they wanted to go and, on the whole, were confident
that they knew how to get there. The old vision is characterized
by Keynes's confidence that 'mankind is solving its economic
problems.'(6) This confidence is exemplified by the conclusion to an
early edition of Samuelson's 'Economics' (1955) written before he
had begun to worry about things like existential vacuums. It
describes the kind of state of which the economist could be proud
to be a citizen:

> We may conclude on a note of profound optimism. The American
> economy is in better shape in the 1950s than it ever was in the
> past. At the present time, possessing only 6 per cent of the
> world's population, it produces some 40 per cent of the world's
> income. And with all its defects, it has behind it a record of the
> most rapid advance of productivity and living standards ever
> achieved anywhere. Our mixed economy - wars aside - has a
> great future before it.

In view of the increasing publicity given to anti-growth arguments, it is easy to forget how very solidly entrenched such
attitudes still are. Economists such as Professor Beckerman,
author of a book entitled 'In Defence of Economic Growth', who
are prepared to defend growth explicitly, may well now be in a
minority among academic economists. But the objective of growth
remains at the core of the economic policies of every major political party in this country, of the Confederation of British
Industries and the trade unions, and is even enshrined, as Article
2, in the Treaty of Rome.

The purposes of the new vision economics are decidedly less
clear. We are told in the declaration that we must not succumb to
gloom and despair. We must not be sucked into the existential
vacuum of hopelessness. We must strive, we are told, for a world
in which it is possible to live with 'dignity', 'hope', and 'justice';
we must have 'a new order of priorities', 'reshaped values' and
'a more humane' vision. But above all we are told - and this is
the only area in which the declarations vision is specific - economists, because they are important and influential people, have a
major role to play in the reordering and reshaping involved in
'the management of our earth home'.

THE ECONOMIST'S PERSPECTIVE

The distinguishing virtue of economics which, according to modern practitioners, entitles economists to such a prominent role in managing the world, is described by David Pearce, one of the sponsors of the Dai Dong Declaration: 'Economics, more than any other discipline, proceeds on the basis of setting out what - to use the jargon - we would call an "objective function" - i.e. saying what we aim to maximize or achieve - and looking at problems in this light.'(7)

The term 'objective function' seems to have become fashionable in the jargon of economics with the development of linear programming, but the ideal of maximizing some unitary index has been around in the subject for some time. It is to be found in Pigou's discussion(8) of the nature and limits of the subject: economics, according to Pigou, was the study of that part of welfare that could be brought into relation with the measuring rod of money. Measured with this rod, that which was biggest and best.

Pigou conceded that economic welfare was not the whole of welfare, and that the boundary between the economic and non-economic was a murky one. He also conceded that on occasion the economic and non-economic could be in conflict. He concluded, however, that as a general rule it was safe to proceed on the presupposition that 'the effect of an economic cause on economic welfare will hold good also for the effect on total welfare'. Further, 'in all cases the burden of proof lies upon those who hold that the presumption should be overruled'. In other words Pigou was saying that economics should proceed on the presumption that God and Mammon are the best of friends and that it is up to those who feel otherwise to disprove it.

This view remains characteristic of up-to-date economics. Keynes elevated the solution of 'the economic problem' to the status of a precondition of improvements in non-economic welfare; only after economic necessity had been eliminated, he argued, could society afford to abandon the measuring rod of money as its standard of progress. Only after Mammon had been appeased could man get on with his 'real, his permanent problem' of how to live 'wisely and agreeably and well'. But most commonly the economic and non-economic have been seen as nothing but the most congenial of companions. This close friendship was viewed by Kaldor as the very foundation of scientific economics.

The statement of Kaldor's, quoted above, was offered as a refutation of Robbins's argument that the scope of that part of economics that could reasonably be called 'scientific' was severely restricted by the impossibility of making interpersonal comparisons of utility(9) - there is, for example, no yardstick that can be used to compare a starving refugee's enjoyment of a bowl of rice with a millionaire's appreciation of a bottle of champagne. This argument between Kaldor and Robbins took place over forty years ago but Kaldor's 'refutation' remains typical of the sleight of mind with which economists still evade the problem that there are at least as many measuring rods of satisfaction as there are people.(10)

The 'scientific' conclusion that Kaldor derived from his basic postulate was set out as follows: 'In all cases, therefore, where a certain policy leads to an increase in physical productivity, and thus of aggregate real income, the economist's case for the policy is quite unaffected by the question of the comparability of individual satisfactions.'(11)

If all satisfactions were capable of being measured by the same monetary yardstick then Kaldor's 'aggregate income' might be, in some sense, 'real'; but they aren't, so it isn't. For the same reason an objective function, with terms purporting to represent the satisfactions and dissatisfactions of a number of different people, is equally unreal. It is the viewing of problems in this unreal light that makes both the old and the new visions characteristically economic.

The preoccupation of economists with material welfare troubled the late Fred Hirsch. He wrote a book about it entitled 'Social Limits to Growth'. Although in it he makes no mention of Samuelson's happiness ratio, it is central to his analysis of society's present predicament. Stripped to its essentials, his argument is that economists, in their endeavours to enlarge the numerator of consumption, have been preoccupied with material consumption and have neglected the consumption of what he calls 'positional goods'. These are goods whose supply no amount of technical ingenuity can increase. They are goods whose appeal lies in their exclusiveness. Jobs that permit people to tell other people what to do (prime ministerships are an example) are goods of extreme scarcity and as a consequence are much sought after. Servants and land are two more examples of goods whose supply technology is incapable of increasing.

The problem then, according to Hirsch, is this: 'The juxtaposition of growth in the material sector and fixity (stagnation without its pejorative connotation) in the positional sector induces a rising trend in the relative price of positional goods' and 'Excess competition in the positional sector has been seen to involve important external costs. If these costs are allowed to become large, a point will come where the damage to society appears too great to justify the individual freedom of action that results in such damage.'

Having identified the problem with some confidence Hirsch becomes rather uncertain about what to do about it: 'The radical aspect of the appropriate solutions for the tensions diagnosed in this book may be precisely their imprecise, general, and evolving form. The prime need is not new instruments but a change in the climate of their use. The radical change needed is to accept that.'

And he goes on to warn his readers that his analysis is of limited use at present for formulating policies because it is 'indeterminate' at two key points. 'The first indeterminacy reflects the lack of a precise criterion for economic efficiency through use of collective action; we do not have a firm grasp on the full implications of collective action, so that the potential inefficiency that can be seen in its omission cannot be firmly categorized as actual ineffi-

ciency or waste. The second indeterminacy resides in the lack of
a quantitative dimension of the critique: it has not been found
possible to estimate over what proportion of economic activity
social limits to growth are in play.'

Hirsch gives principal credit for stimulating his insights into
the importance of positional goods to essays by two economist
predecessors, Harrod in 1958, and Wicksteed in 1910. Beyond the
pale of economics the idea might well go back even farther. The
three favourite games in my 3-year-old daughter's play-group at
the present time are all organized around the theme of positional
goods. They are 'King of the Castle', 'Hoarding all the Toys' and
'Bullying the Little Ones' - games that have been played ever
since Adam and Eve left Eden. And, more recently, surely R.L.
Stevenson deserves a mention for recognizing, in its embryonic
form, the avidity of homunculus economicus for positional goods:

> When I am grown to man's estate
> I shall be very proud and great,
> And tell the other girls and boys
> Not to meddle with my toys.

Most parents know what policy to follow when the competition in
positional games becomes excessive and leads to important costs -
that is, when it ends in tears. They try to interest the children
in non-competitive activities, perhaps singing and dancing, in
which pleasure can be taken from the pleasure of others, and not
at the expense of others. And they do not seem to need a firm
quantitative grasp on the full implications of collective action before
intervening in a play-group riot.

Another economist whose writings manifest an ambivalence towards
the materialism of his discipline is Ezra Mishan. In 'The Economic
Growth Debate', published in 1977, although Mishan still calls
himself an economist, he has almost broken free of economics. His
struggle to break free has been a hard one and demonstrates the
strength of the mental shackles with which economic practitioners
are bound.

In 1967 he published a curious but influential book called 'The
Costs of Economic Growth'. It was curious because the argument
against economic growth was confined to what he called a 'digression' on the unmeasurable costs of economic growth. He called it
a digression because it dealt with considerations that he argued
lay beyond formal economic analysis. Perhaps the most persuasive
section of this part of the book was his attack on 'the cult of
efficiency'. 'Today', he complained, 'no refuge remains from the
desperate universal clamour for more efficiency . But the central
argument of the main body of the text was that conventional
economics was an inefficient custodian of the public weal. It was inefficient, he argued, because it left out of its calculations 'externalities', which could not, in the present state of the art, be
estimated by the measuring rod of money. He admonished:

> If the problem [of externalities] is to be tackled by society, the
> economist must persist in revealing the nature of the beast, and

must suggest the circumstances under which meaningful magnitudes may be attributed to external effects. Nor should he shirk detailed description of cases wherever the social consequences that escape the pricing system appear to be so involved that a comprehensive criterion for evaluating them cannot, as yet, be satisfactorily evolved.

'As yet' - he took his admonition seriously. Five years later he published 'Cost-Benefit Analysis' in which he makes clear that his objection is not to economic growth, but to economic growth inefficiently calculated.

Economic growth comprises both population growth and growth of per capita real income. Together they contribute to the growth of benefits over time arising from any investment today. The faster the rate of economic growth, then the swifter, in general, will be the rate of growth of future benefits. In cost-benefit studies the likelihood of such growth-induced benefits has to be taken into account.

His determination to capture all such growth-induced benefits within the framework of his analysis led him, in Chapters 22 and 23 of the book, to attempt the cash valuation of life itself.

In his most recent book the 'digression' of 1967 has grown into the main theme, and the formal discussion of economics has been relegated to the status of a digression. After introducing himself as an economist and doing a few steps around the externalities theme, he announces that he is 'writing to doff my professional cap and to speak without the authority of economic analysis - to appeal to imagination, reason and good sense, only'.

He is almost free, but not yet. He is an anti-growth economist, a flesh-and-blood oxymoron. Economics is cash-quantified egoism. The numbers in which it deals are derived by applying the measuring rod of money to the behaviour of profit maximizers. Economic analysis presumes profits to be good and losses to be bad. Without this presumption it has no point. There may well be profits and losses that, as yet, economists have not succeeded in measuring very precisely but, in principle, by definition, the growth of profit is good.

The anti-growth economist is caught in a cleft stick. His conceptual heritage consists of mathematical models that maximize or minimize monetary indices. Having concluded, rightly, that these indices, as they have been calculated traditionally, are useless as measures of welfare, his solution is to calculate them differently. For the anti-growth economist, man remains a utility maximizer; he simply must be induced to revalue the terms in his objective function. The index to be optimized must include not only material goods but spiritual, aesthetic and positional goods as well. Not only clean air and fresh water but things like beauty, romance, security, contentment, community spirit, and a sense of personal significance are externalities that must be included in the function. Formulating the problems of the world in this way permits the economist to bring all the old conceptual machinery of orthodox economics to bear upon them. He can reorder consumption patterns

by applying appropriate taxes and subsidies; by raising the price
of material (polluting) goods and lowering the price of non-
material (non-polluting) goods he can divert consumption into environmentally desirable channels. It also gets round the awkward
problem of having to advocate retrenchment. 'Growth' and 'Progress' are still perfectly possible; they are simply redefined.

In 'The Economic Growth Debate' Mishan appears to have become reconciled to the idea that some profits and losses may be
so involved that they may never yield to the measuring rod of
money. And his message, quite eloquently argued, is that the unmeasurable losses of economic growth far outweight both the
measurable and unmeasurable profits. He itemizes the losses and
it makes an impressive list. It includes food, shelter, nature,
leisure, instinctual enjoyment, love, trust, self-esteem, kith and
kin, customs and mores, roles and place, the moral code, the
great myths, and personal freedom. His survey leaves him in a
state of helpless despair, aggravated by a case of purple prose.
'It is not possible, then, to end on a note of even qualified optimism. . . . Western civilization, the civilization born of high
hopes and auspicious heralding [is] today frothing with power
and glee - and being piped gaily to the brink of the abyss. And
all that yet might stay the fatal plunge lying in the mud, dis-
carded and in decay.'

The anti-growth economist is indeed helpless; he can offer
nothing to fill 'the existential vacuum'. Does he set out to make
society's books show a loss? Anti-growth economics rests upon a
contradiction. The meaning and purpose of life cannot be reduced
to an objective function because spiritual and aesthetic 'goods'
do not have cash equivalents. They are antithetical to the very
idea of cash equivalents. The only way an anti-growth economist
can give them a cash value is by relating them in a quite arbitrary
way to the material goods whose consumption he proposes to discourage. The only incentives he can employ in order to induce
people to become less materialistic are materialistic incentives. For
these to work they depend upon the very attitudes they seek to
alter. If a wealthy industrialist should take his advice and give
up his wicked acquisitive ways to spend his life cultivating his
garden and being friendly to his neighbours, how should this be
recorded? If it is a social profit, how should it be measured? The
best the economist can do with his monetary yardstick is to value
the industrialist's new circumstances by reference to the market
value to the things he has left behind. If he values all his abandoned material goods and positional chattels and adds them up he
can produce an estimate of the opportunity cost of being a neigh-
bourly gardener. But what could such a number mean that values
what is deemed good in terms of what is deemed bad?

Keynes, the most famous pro-growth economist of them all, was
greatly impressed by his own store of positional goods. He suggested condescendingly, and unconvincingly, that they had been
pressed upon him and he would be happy to be rid of them: 'If
economists could manage to get themselves thought of as humble,

competent people, on a level with dentists, that would be splendid!'

Good advice, its suspect sincerity notwithstanding. But the principal enemy is not the pro-growth economist, whom Mishan lashes remorselessly throughout his book, but the patron of all economists, homunculus economicus. Keynes's dream of wealth beyond the ambition of avarice is as old as Mammon. And most people have always known that no amount of science and technology and clever economic calculation could ever make it come true.

## REFERENCES AND NOTES

1   Kaldor, N. (1939), Welfare Propositions of Economics and Interpersonal Comparisons of Utility, 'Economics Journal', 49, pp. 549-52.
2   Samuelson, P. (1955), 'Economics', McGraw-Hill, p. 707.
3   Keynes, J.M. (1930), Economic Possibilities for our Grandchildren, in 'Essays in Persuasion', Rupert Hart-Davies, London, 1952 ed.
4   Lenin, N., The State and Revolution, Selected Works, reprinted in 'The Marxists', C. Wright Mills (ed.), Pelican, 1962.
5   'Dai Dong: Towards a Human Economics', a declaration/ petition circulated to economists around the world in 1974-5.
6   Keynes (1930), op.cit.
7   Pearce, D. (1973), letter to the editor, 'Ecologist', December.
8   Pigou, A.C. (1920), 'The Economics of Welfare', Macmillan, 3rd ed. 1929.
9   Robbins, L. (1938), Interpersonal Comparisons of Utility: A Comment, 'Economics Journal', vol. 48, pp. 635-41.
10  Robbins, having won the argument, immediately handed it back to Kaldor; in order to salvage economics he agreed to assume that which he had demonstrated he could not prove:
     I am distressed that anything that I have said should give rise to recurrent dispute which suggests to the outside world a disunity among economists which I am persuaded does not exist. . . . They [his critics] think that propositions based upon the assumption of equality are essentially part of economic science. I think that the assumption of equality comes from outside and that its justification is more ethical than scientific. But we all agree that it is fitting that such assumptions should be made and their implications explored with the aid of the economist's technique.
11  Kaldor (1939), op.cit.

# 16 TRANSPORTATION FOR LUDDITES

At the beginning of the nineteenth century machines were dramatically transforming economic and social relations. They still are. But the pace of change is now much faster. In the realm of transport, what is at stake in the debate about the consequences of mechanical progress is nothing less than a way of life. At the beginning of the present century, Britain was still a predominantly pedestrian society. The changes in the country's way of life brought about by the great increase in mechanized mobility during this century have been profound. The further increase in mobility during the remainder of the century that is anticipated by the government's forecasters, and fostered by its transport policies, is about as great as that which has taken place since 1900. The further changes in the country's way of life that would take place as a consequence of such an increase can reasonably be presumed to be equally profound.

Such transformations inevitably benefit some people at the expense of others. Certain occupations and areas fall by the economic wayside while new occupations and areas rise to prominence. Village and neighbourhood shops and small urban industries have disappeared in large numbers. Villages and the inner areas of cities have declined while the suburbs have prospered. Drovers, teamsters and wheelwrights are no more; they have been superseded by lorry and train drivers, pilots and assembly-line workers.

Economic development is an evolutionary process in which only the economically fit survive and prosper, and Social Darwinists since before the time of Darwin have said 'so be it'. The rate of the process is still increasing. It has been estimated that every ten years in the United States 57 per cent of all industrial jobs are lost through failure to survive competition with producers who are more efficient or more closely attuned to consumer fashions.(1) This, economist Walter Eltis insists, is wholly desirable: 'Growth is achieved through the continuing destruction of bad jobs and their replacement with good ones'.(2) The fact that the jobs that are destroyed and those that are created are rarely in the same place, and are frequently separated by a distance of hundreds or even thousands of kilometres, is dismissed by Eltis with an insouciance common to those who deal with welfare in the abstract; it is, he says, 'merely a complication'. All those who would allow such complications to stand in the path of economic growth are, he declares, 'irretrievably Luddite'.

Darwinists are characteristically possessed of a strong sense of personal fitness. This is not surprising, because one's view of

whether or not a social transformation is desirable will usually be
strongly influenced by whether one expects to gain or lose by it.
Whether one prefers to be thought irretrievably Luddite will depend on whether, in contemplating the changes impending in one's
own life, one sees instructive analogies in the predicament of the
original Luddites and feels sympathy with their response.

An ironic appreciation of the Luddite's predicament is found in
'Hansard'(3) which records Lord Byron's contribution to the debate
in the House of Lords in 1812 on a Bill proposing 'more exemplary
punishment' (i.e. hanging in place of transportation) for the
offence of destroying or injuring machinery used in the stocking
industry:

> the police, however useless, were by no means idle: several
> notorious delinquents had been detected; men, liable to convic-
> tion, on the clearest evidence, of the capital crime of Poverty;
> men, who had been nefariously guilty of lawfully begetting
> several children, whom, thanks to the times! they were unable
> to maintain. Considerable injury has been done to the proprie-
> tors of the improved Frames. These machines were to them an
> advantage, in as much as they superseded the necessity of employ-
> ing a number of workmen, who were left in consequence to
> starve. By the adoption of one species of Frame in particular,
> one man performed the work of many, and the superfluous
> labourers were thrown out of employment. Yet it is to be obser-
> ved, that the work thus executed was inferior in quality; not
> marketable at home, and merely hurried over with a view to
> exportation. It was called in the cant of the trade, by the name
> of 'Spider work'. The rejected workmen in the blindness of their
> ignorance, instead of rejoicing at these improvements in arts
> so beneficial to mankind, conceived themselves to be sacrificed
> to improvements in mechanism. In the foolishness of their hearts
> they imagined, that the maintenance and well doing of the in-
> dustrious poor, were objects of greater consequence than the
> enrichment of a few individuals by any improvement, in the
> implements of trade, which threw the workmen out of employ-
> ment and rendered the labourer unworthy of his hire. And it
> must be confessed that although the adoption of the enlarged
> machinery in that state of our commerce which the country once
> boasted, might have been beneficial to the master without being
> detrimental to the servant; yet in the present situation of our
> manufactures, rotting in warehouses, without a prospect of
> exportation with the demand for work and workmen equally
> diminished; Frames of this description, tend materially to
> aggravate the distress and discontent of the disappointed
> sufferers.

Following Byron in the debate was Lord Lauderdale, a noted economist of his day and author of an 'Inquiry into the Nature and
Origin of Public Wealth'. He presented a less complicated analysis:

> the outrages in Nottingham originated in a mistaken notion of
> those concerned in them, of their own interests, for nothing
> could be more certain than that every improvement of machinery

contributed to improve the conditions of persons employed in
the manufactures in which such improvements were made, there
being in a very short time after such improvements were introduced a greater demand for labour than there was before.
It is a debate with a remarkably modern flavour. The same issues
are still being debated. It might be argued that the superior
material circumstances in which today's debate is taking place are
a vindication, of sorts, of Lord Lauderdale's position. From the
beginnings of the industrial revolution it seems Luddites and other
kindred spirits have had to be dragged kicking and screaming
down the road of progress. They may not have enjoyed the journey,
but travel, after all, is a means to an end, and the stage reached
by Britain today in the last quarter of the twentieth century is
clearly preferable to the journeyers' pre-industrial starting point.
Thus, all those who kick and scream today in protest at society's
still increasing dependence on machinery, who resist new airports
and nuclear power-stations, who protest against new roads, and
bigger lorries, and hypermarkets, and supersonic airliners, and
the computerization of commerce and industry and government –
all who metaphorically, and sometimes physically, stand in the
path of the bulldozers of technological progress – are Luddites
with 'a mistaken notion of their own interests'.
   All this is very well trodden ground. The debate about the
standard of living in Britain since the beginning of the Industrial
Revolution is a staple of economics, history and sociology. It is a
debate that has been generalized to a debate about the nature of
economic and social development  everywhere, to a debate about
the definition of progress. It is a field upon which the intellectual
Goliaths and not a few Davids of the nineteenth and twentieth
centuries have done bloody, inconclusive battle; and it is a field
where ignorant armies of the late twentieth century still clash by
night.
   Right, left and centre the field is dominated by the great
materialist armies. Although there is fierce skirmishing along the
flanks of their columns they are all driving in the same direction,
towards what Rostow called without apparent ironic intention 'the
stage of high mass consumption' towards the Keynesian 'destination
of economic bliss' towards the Marxist secular heaven where 'all
the springs of cooperative wealth flow more abundantly'. In each
of these armies the generals are economists and the quartermasters
are the transport planners. The economic strategies of present-day
governments of all ideologies see transport as the means by which
virtually all activities of economic consequence are related. The
flow of wealth is the flow of traffic, and it is the quartermasters'
job to make it flow more freely and abundantly.
   But what is wrong with such a goal? Can it seriously be argued
that life for the affluent members of industrialized countries today
is not more comfortable in virtually every material respect than it
was in 1812? Or is there any evidence of any sort of spiritual or
intellectual superiority in those who lived in the harder times of
the early nineteenth century? The matter of fact tone in which

parliamentarians in 1812 debated whether or not to hang Luddites who protested against their material wretchedness by breaking machinery suggests not. And if not, why should we not strive to maintain the progress achieved in the past 165 years? While writing this I can pause and observe a group of bricklayers at work outside my window. The council houses they are building are superior to anything the common man dreamed of in 1812 and their working conditions, while obviously physically demanding, are also a vast improvement on the lot of their nineteenth-century predecessors. One might prefer life in one materialistic army to life in another, but surely to dispute the desirability of the direction in which they are marching is an indulgence worthy of Marie Antionette.

Such appeals to history as a guide to what we ought to do in the future are specious. The fact that someone prefers the lot that history has provided him, to the alternative it might have provided had there been no industrial or transport revolutions, might lead him to bless his good luck, but it ought not to lead him to an uncritical acceptance of the means by which this lot was delivered. The present philanthropic generation of Rockefellers might well enjoy its privileged position in life, but that does not mean that the economic buccaneering of the founder of the family fortune should serve as our model for the future conduct of economic affairs. Nor does the affluent good fortune of the middle classes of the industrial world today morally justify the historical catalogue of misery and exploitation that produced its wealth, or the continuing exploitation of the poor world, and the extravagant consumption of posterity's non-renewable resources that help to sustain it today. Nor does it provide a justification for the pursuit of economic growth for the indefinite future.


THE INDEFINITE FUTURE

In 1973 'The Sunday Times' published a map(4) showing the way in which Concorde would shrink the world. It consisted of two maps of the world, one drawn at half the scale of the other, with the smaller map superimposed on the larger. Both maps were centred on Paris. 'The point' of the composite map was 'that it would only take a few hours longer to fly [in Concorde] from Europe to most important parts of the world than to fly the Atlantic in a Jumbo Jet'.

The map was of course a misleading oversimplification. The distance transformation employed in drawing it was based on a direct comparison of the cruising speed of a Concorde with that of a Jumbo Jet. It ignored the fact of Concorde's limited range, which would require time-consuming detours and fuelling stops, and assumed away the question of overland 'booming rights' which remained, and still remain, to be negotiated. But Terry Hughes, the author of the article that accompanied the map, clearly thought the map not only true but an extremely effective piece of pro-

Concorde propaganda. That shrinking was good for the world he had no doubt. That his readers would agree that shrinking was good for the world he also did not appear to doubt. Certainly he did not feel that it was necessary to explain to them why it was a good thing. It was sufficient simply to demonstrate that Concorde would do it.

His assessment of the general reader's reaction was probably fairly accurate. Certainly market researchers, advertising copy-writers, and other monitors and manipulators of the popular psyche have arrived at similar assessments. It is sufficient for British Airways to call their tickets 'Earth Shrinkers' to be confident of selling more of them. It is sufficient for British Rail to demonstrate that electrification has shrunk Britain to convince the public that it is in the vanguard of progress, and sufficient to boast of the coming wonders of the 'Advanced Passenger Train' to prove that it intends to remain there. It is sufficient to note that the new M4 motorway has brought South Wales closer to London to silence the project's detractors. And it is sufficient for the Post Office to remind people that the telephone brings them closer together to show that it is a wholly beneficial social institution.

Who then could doubt that this shrinkage really is progress? Of course the technologists of mobility agree that we must try to reduce the fumes and noise, and limit the numbers of homes des-troyed and neighbourhoods disrupted, that we must fly super-sonically only over the sea or 'sparsely populated areas', and that we must bury the wires and cables, and landscape motorways artistically. We must, of course, seek to minimize the undesirable by-products of increased mobility. But increased mobility, itself, is indisputably a benefit. Or is it?

What precisely are the benefits of increased mobility? And is there some point at which diminishing returns set in, or will increasing mobility represent progress for ever? Such simple questions receive surprisingly evasive answers. 'Increased mo-bility promotes economic growth' is one confident answer that leads nowhere; it is no longer possible simply to appeal to a grow-ing gross national product as some ultimate sanction for a point of view. The very same questions are being asked of it. 'It helps the balance of payments' and 'it creates employment' are two more answers. For ever? If current growth rates must level off some time, will not the balance of payments and employment problems associated with the levelling off be greater the longer it is de-layed? 'They said the same thing about railways . . . cars . . . telephones . . .' or 'What about the Luddites?' are curious rhetorical devices adduced as argument. The economic argument for ever-increasing mobility twists and turns, but ultimately it either dissolves or ends up at the fundamental, dogmatic, and old-fashioned assertion that ever-increasing economic growth in a finite world is both possible and good.

The ground then shifts. There are psychological and social arguments to be considered. Lord Beswick(5) a prominent Labour Party spokesman on aviation matters, and now head of British

Aerospace, puts the following case:

It is inescapable that in any progressive industrial society there must be a pioneering spearhead technology. It is not simply a matter of establishing the facts, or of finding out what can or cannot be done. There is a psychological spin-off, a constructive feeling of pride, a stimulating sense of prestige, if one's own society can claim to lead in any given field. Concorde, the RB 211 and Harrier can reasonably be said to afford this constructive stimulation.

This argument is part chauvinism, part truism, and part whistling in the dark. It is no doubt true that pride and prestige are conducive to pleasurable and stimulating psychological states, but these states are crucially dependent upon public recognition of one's 'achievement'. There are many 'fields' whose leadership no one can take pride in. In the absence of any criteria for distinguishing technological achievements from technological follies, 'psychological spin-off' is likely to be very difficult to sustain. Such criteria cannot be found within the tautology that asserts that something is good because it makes one proud because it is good.


## A STIMULATING SENSE OR PRESTIGE

It has already been noted in Chapter 6 that one's vision of the world is related to the speed and height at which one travels. Concorde provides abundant evidence that the low level of resolution associated with high speeds and altitudes does indeed produce a stimulating sense of prestige. On 22 January 1974, in the immediate aftermath of the energy crisis, the House of Lords discussed Concorde, the aircraft that consumes more energy per passenger kilometre than any other with the exception of the helicopter. Lord Beswick(6) asked his peers 'Is it not the fact that Concorde is the only aircraft in the world that has been designed to carry essential travellers; and ought we not now, with the shortage of fuel supplies, so allocate our fuel supplies that we use it for essential purposes?' On behalf of the Conservative government Lord Aberdare(7) replied: 'My Lords, I thank the noble Lord; he has put an opinion with which I thoroughly agree.'

In May 1976, Britain's Royal Automobile Club, in recognition of Concorde's splendid contribution to 'public transport', presented its makers with its prestigious Diamond Jubilee Trophy. Upon inspection of the list of those attending the banquet at which the award was presented, headed by Lord Mountbatten, one might plausibly speculate that they were all 'essential travellers', and that their essential duties had not required them to travel by bus in years, if ever.

One of the difficulties with supersonic public transport is that it trails a boom carpet in its wake at least 80 kilometres wide. Among advocates of the plane there are two schools of thought about how to deal with the problem. One is to argue that the inessential people 16 kilometres below do not matter. Mr. Charles Gardner, publicity manager for the British Aircraft Corporation, belongs to this school. Explaining why it would be acceptable to fly supersonically over Australia he said 'In the Australian desert there's nothing except a couple of abos and lots of kangaroos'.(8) The second is to argue that the sound of the Global Villagers going about their business supersonically is positively enjoyable. Sir James Lighthill, director of the Royal Aircraft Establishment at the time when crucial design work on Concorde was being done there in the early 1960s, belongs to this school:

> Concorde is found to have a very friendly boom, 'like a pair of gleeful handclaps'. . . . I conclude that public acceptance of Concorde's friendly boom can be confidently expected in the longer term, when the public have got to know its nature for themselves, and at the same time have come to appreciate the human advantages from all long-distance flight times being more than halved.(9)

In Lebanon, at the time of writing, while essential travellers were celebrating their essentialness overhead with free champagne and gleeful double handclaps, newspapers were reporting numerous cases of damage to houses beneath Concorde's flight path. A north Lebanon Member of Parliament was denouncing the 'colonialism of the air' of Britain and France, and a parliamentary committee was pressing the government to ban supersonic overflights.(10) In London, Press spokesmen for neither the Lebanese nor British governments could explain the nature of the inducements that had led the Lebanese government to inflict on its poeple something that the British and French governments declined to inflict on their people. It remains unexplained why Lebanon's government, which doubtless includes many essential travellers, should be more appreciative of the human advantages of halving long-distance flight times than their counterparts in the countries that created the plane, especially since Concorde does not even stop in Lebanon.


## PSYCHOLOGICAL SPIN-OFF

Mobility brings with it power and significance. We were told in the 'Sunday Times' article, quoted above, that Concorde would permit Europeans to reach 'most important parts of the world' very quickly. These 'important parts' of the world were not specified but they are easy to guess; they are the cities with busy international airports, the cosmopolitan capital cities of the world; they are the expensive, upper-class neighbourhoods of the Global Village. The rest of the world becomes less important to the increasingly mobile few; the people who live there become too small to be seen individually and can enter the global decision-makers' thoughts only in the form of statistical abstractions. They become subservient 'factors' in the calculation of global 'strategies'. Not surprisingly the less mobile often fail to appreciate the grand designs of the Global Villagers whose 'shrunken' world have subsumed their own. They frequently do not want the airports and motorways that speed important people about their business. But their lack of appreciation is easily derided as provincial insularity and narrowness of vision.

Being a Global Villager can be satisfying work. Taking decisions that affect the lives of large numbers of people confirms one's own significance. But having one's life arbitrarily controlled by others is alienating. In order to sustain the self-esteem of the controller, and to minimize the alienation of the controlled, it is therefore necessary to devise an appropriate disguise for large-scale decision-taking operations. Consider the decision-taking exercise of the Roskill Commission. That some people were alienated by its decision is obvious; farmers, shopkeepers, neighbours - people recognizable as individuals, real people whose pictures appeared in newspapers - were sufficiently incensed to burn an effigy of Justice Roskill. It was the job of the Roskill Commission to show that this opposition, while certainly understandable, was narrow and selfish when placed in the context of the larger good that would be served by a new airport. But who were the people whose interests outweighed those of the airport's opponents? Where were the individuals whose desire to travel was so important? Nowhere. They were a statistical extrapolation; they were a graph of traffic rising to 300 million by the year 2006; they were an abstraction that could not be cross-examined. Who were their spokesmen? They were Global Villagers from the British Airports Authority, the airlines, and the tourist industry. They were people in the mobility business whose 'job satisfaction' depends upon public recognition of their achievements. They were, and are, the motive force behind Lord Beswick's 'spearhead technology'. The graph that they presented to the tribunal (of important and mobile men who were to decide the airport controversy) was little more than an extrapolation of presumed public acclaim to the year 2006; the 'stimulating sense of prestige' that they accept as their due for providing the facilities to carry the existing volume of traffic, would be as nothing compared to the recognition that they presumed would be their due if they could increase this volume of traffic fifteenfold.

No doubt many 'ordinary' people can be found to agree that new airports and motorways are a good thing. Although much of the recognition in which the transport technologists bask is simply their own reflected propaganda, their pride is not entirely self-induced. People do recognize an equation between mobility on the one hand and freedom and power on the other, and reason that if only they had more of the one they would have more of the others. But it is an equation that ignores the relative nature of mobility, freedom and power, and that ignores the relationship between the general level of mobility in a society, and its scale. Large-scale societies have no more people at the top than small-scale societies, only more at the bottom. Thoreau, who lived and wrote in the nineteenth century when the railway building booms in England and America were reaching their peaks, saw very

clearly where the boundless enthusiasms of the transport technologists were leading.

To make a railroad round the world available to all mankind is
equivalent to grading the whole surface of the planet. Men have
an indistinct notion that if they keep up this activity of joint
stocks and spades long enough all will at length ride somewhere,
in next to no time and for nothing; but though a crowd rushes
to the depot, and the conductor shouts 'All aboard!' when the
smoke is blown away and the vapour condensed, it will be perceived that a few are riding, but the rest are run over.(11)

## CONTROLLING THE KNOWN WORLD

Developments in transport and communications technology have
already necessitated many social controls. Speed limits, blood
alcohol limits, driving tests and licences, parking restrictions,
laws to regulate noise and fumes, laws permitting the compulsory
purchase of land for transport developments, compulsory vaccina-
tion, and the vast apparatus of customs and immigration are only
a few of the most obvious social controls made necessary by increasing mobility. But even the impressive array of existing controls is generally deemed inadequate; the carnage on the motorways, the congestion in cities, the noise around airports, and a
whole range of indices of pollution and social discord are all
generally considered to be above acceptable levels and in need of
control.

Further, increased mobility is creating new problems of control
that are unprecedented in scale and kind. As traffic across traditional political and administrative boundaries increases, the
politicians and administrators progressively lose control over
matters that they have traditionally controlled. The highly mobile
commuter, for example, can make his money in one authority and
spend it in another. Or, the efforts of local authorities to relieve
the lot of their poor can be frustrated by an influx of poor migrants who are attracted by these efforts. On a larger scale, the
internal accounting arrangements of large international corporations result in large international transfers of wealth; financiers
move offshore to evade the laws of nations that they find uncon-
genial; and international shipping companies pollute international
waters with impunity.

But increasing mobility is not only creating a need for more and
larger social controls, it is weakening many existing controls. The
greater mobility in daily life is reflected in greater residential
mobility. An extreme manifestation of this kind of mobility is cited
by Toffler(12); over half the 885 000 listings in the Washington DC
telephone directory in 1969 were different from the listings of the
year before. Although telephones enable people to keep in touch
after they have moved apart, and improved transport facilitates
family reunions, such intermittent contact at best sustains much
weaker relationships than those that have been disrupted. To the

extent that families and neighbourhoods act as agents of social control, such controls are weakened by greater mobility. Whether the net result of such changes is 'liberating' or not depends very greatly on the nature of the controls by which they are replaced.

## FREEDOM AND CONTROL

'Our wills . . . have just so much power as God willed and fore-knew that they should have; and therefore whatever power they have, they have it within most certain limits'. (St Augustine (13))

Thus did Saint Augustine reconcile his belief in an omnipotent deity with free will. Although it is a rather unsatisfactory recon-ciliation, it provides a very good description of the nature of human freedom. A man is constrained in a great variety of ways; he is constrained both by his physiological needs and limitations and by his physical environment. He is also constrained socially by the values he holds in common with other men. When he has a felt need to live by these values, he calls them attitudes or beliefs and does not generally recognize them as constraints, but when they are formerly codified as laws and enforced by social institutions, their constraining nature is more readily re-cognized.

The two sets of constraints, physical and social, complement each other in ways that a biologist would call symbiotic. For example, where advances in medicine have succeeded in removing important physical constraints on population growth their replace-ment by social constraints is essential to avert catastrophic demo-graphic imbalances. In societies where the impact of modern medicine has been recent and sudden, the replacement of old and well established social constraints, that is attitudes towards large families and the taking and prolonging of life, by new constraints that are 'ecologically more appropriate', is clearly a very painful process. The new constraints, perhaps because they are new, have not been assimilated as attitudes, and are resisted as callous restrictions of fundamental human rights.

It is a basic premise of this book, and indeed of all polemical writing, including that of professed determinists such as Marx, that within the constraints that confine him, man is free to act as he will; a pure determinist can have nothing to say in debates about what ought to be. Man, in most cultures and certainly in our own, has always utilized his limited powers and freedom of manoeuvre in an attempt to push back the constraints that re-strain him in order to enlarge the area of freedom within. It is a technocratic article of faith that an increasing mastery on his material circumstances will automatically give man greater freedom.

This faith in technology has prevailed throughout the 'developed world'; it has been the very foundation of this world's develop-ment. Although many people have become alarmed by the evidence of the unanticipated side effects of our technological achievements, the true believer is not shaken by such evidence. He is possessed

of an open-ended optimism; rather than admit that man is not
perfectible by science, he argues that the problem is simply one
of imbalance. For too long, he says, technological ingenuity has
been focused on non-human nature; the balance, it is argued,
must be, and can be, restored by shifting the focus of science to
social problems. Nobel prize winning physicist Denis Gabor puts
it this way:

> The problem lies not in technology but in the fact that man is
> not prepared for it.
>
> What science can do is solve the extremely difficult problem
> of creating the right conditions for the development of culture;
> this should enable us to keep the maximum possible individual
> freedom.
>
> For the Apollo Moon programme about six hundred thousand
> people worked together as one team and made the project run
> with split-second accuracy. No wonder these people are very
> proud of their achievement and say; 'Let us now apply our
> methods to social problems.' I am quite convinced that by tak-
> ing an equally gifted set of people - even taking the same
> engineers, with, of course, a sprinkling of economists and
> social scientists and the like - and giving them social problems
> to solve, such as for instance the race problem in the United
> States, the social integration of the American cities, the build-
> ing of new cities with adequate urban transport etc., that all
> these problems could be solved, because these engineers and
> scientists have evolved effective methods of integrated plan-
> ning and they have an absolutely wonderful system of co-
> operation. Once a dream becomes a project, the engineers can
> deal with it!(14)

Another believer in the capacities of scientists and engineers to
solve social problems is G.T. Seaborg, former chairman of the
United States Atomic Energy Commission. Here is his vision of the
Utopia he believes them capable of creating. He calls it 'The
Erehwon Machine'.(15)

> Operating on natural, self-renewing sources of energy, it gen-
> erates power with an efficiency approaching 100 per cent and
> still makes beneficial use of its minute amount of waste heat.
> Anticipating public life-styles, it provides durable goods and
> services equitably to all people. It recalls all waste, reduces
> the waste to its elemental form, then recombines it as basic
> materials with needed raw resources that the machine extracts
> from nature with a negligible environmental impact.
>
> The Erehwon Machine is self-adjusting to a changing labour
> market; it provides retraining to workers as it upgrades itself
> and administers welfare to all it cannot employ. To the unem-
> ployed and unemployable it offers satisfying leisure activities
> and psychological reorientation to a non-work ethic. And in
> doing all this for man and nature, it still shows a profit to its
> stockholders and a favourable effect on the international balance
> of payments.

Although Seaborg concedes that all this is a bit much to hope for he insists 'Seriously, I think much if not all of the tasks I have assigned to the'Erehwon Machine is what most of us would like to take place.'

The man who has come closest to putting such ideas into practice is Stafford Beer. Like Gabor and Seaborg he is an eminence within his own scientific discipline, he is a former president of the Operational Research Society, and also professes a concern for individual freedom. In the 1970s he was engaged by President Allende of Chile to build and install what Professor Beer calls a 'Liberty Machine'. President Allende was overthrown before the Machine had become fully operational and it is not known the extent to which his successors have succeeded in implementing it. This is how Professor Beer describes it:

> The real-time control system I have so briefly described is founded on the following elements: a cybernetic model of any viable system; a cybernetic analysis of the real-life systems appropriate to each level of recursion, and their iconic representation; a design of a large number of interlocking homeostats; the provision of a national communications network capable of operating now on a daily basis and eventually on the basis of continuous input; variety engineering throughout the system to incorporate filtration on the human brain's scale; and the Cyberstride computer program suite capable of monitoring inputs, indexical calculations, taxonomic regulation, short-term forecasting by Bayesian probability theory, autonomic exception reporting, and algedonic feedback. It makes quite a package, and it exists. It represents a system of here-and-now management of the economy that is not based on historical records, but on an immediate awareness of the state of affairs and the projection of that awareness into the short term future.

The nerve centre of this system of control is the operations room. Again we will let Professor Beer describe it.

> If the connotation of that phrase[operations room] reminds some people of a wartime headquarters, the allusion is quite deliberate. For in the opsroom real-time information is laid out, quite graphically, for immediate decision; and in the opsroom a synoptic view of the whole battle is made plain, so that the total system can be encompassed by human powers of foresight. We used every scrap of relevant scientific knowledge of brain processes, knowledge from applied and group psychology, knowledge from ergonomics.
>
> The opsroom looks like a film set for a futuristic film. But it is not science fiction; it is science fact. It exists, and it works; it exists and it works for the worker as well as the minister.(16)

For those puzzled to know what all this has to do with Liberty, Professor Beer explains that Liberty is a new technical term in the jargon of operational research: 'Liberty may indeed be usefully redefined for our current technological era. It [sic] would say that *competent information is free to act* - and that this is the principle on which the new Liberty Machine should be designed.'(17)

The objective of social engineers, just as of physical scientists and engineers, is to understand the world in order to exert greater control over it. The one absolutely crucial distinction between the objectives of the physical and social sciences is that the one aspires to control the physical world and the other aspires to control the social; the one would control *it*, the other would control *us*.

Although this objective, when formulated so starkly, would be denied by many social scientists, it can be shown to be implicit in their work; they perform impressive acrobatics pretending that it is not. They describe future Utopias that are controlled without controllers: they speak of 'self-regulating societies', 'stable ecological systems', and states of 'social homeostasis' which are internally controlled by impersonal 'viable governors'. But this sort of language begs an important question.

How are we to get to one of these stable Utopian states from our present state of social chaos without someone imposing direct and Draconian controls? The answer would appear to be that we cannot. Edward Goldsmith,(18) the editor of the 'Ecologist', when he descends from the abstract ecological principles of his 'Blueprint for Survival' to a specific problem, readily admits the need for controls of a very direct and nasty kind. Speaking of the population problem he says 'we must introduce the correct cultural controls to prevent them from wanting so many children. It is a sort of cultural engineering and I believe it could be done.' His separation of 'we' and 'them' is in the best tradition of scientific detachment; but, in the absence of a precise definition, the dichotomy raises very real anxieties among those who suspect that they might be allocated to the less desirable side of it.

Professor Beer is quite clear about whom he thinks out to do the allocating. 'How', he asks, 'can we say that we know best and everyone else is wrong?' His answer is characteristic of all those who possess what he calls 'competent information'.

It is that we are *responsible*. We are not responsible because we have been elected to govern affairs; we are responsible because cybernetics, that science of effective organization, is our profession. Such understanding of this subject as there is, we have. Therefore we must speak out. . . . We do not know as much as we would like to know before making such a stand as is here proposed. . . . We must use such tools as we have, and use them now. . . . Some who are here, and some great men and wonderful friends now dead, guessed that the tool kit would not be finished in time. Today we know that the moment has come for us to start, and we must do the best we can.(19)

## A CHOICE OF CONTROLS

It appears that society has, within very severe limits, a choice in the constraints that give it shape and order. But we cannot choose to have no constraints at all. The most notable consequence

of the success of the transport and communications revolution in
diminishing the physical constraint of man's immobility, has been
the development of a variety of social constraints to take its
place. An imaginative extrapolation of this revolution and its consequences produces a very bleak scenario. The further the scientist pushes back man's physical constraints, the clearer and more
depressing becomes man's view of infinity, the more the significance of man is diminished, and the greater becomes his need for
the psychological sustenance that can only be provided by satisfying relations with his fellow men. But the larger the scale of
society grows, the less capable it becomes of recognizing the individuals of which it is comprised. The controls required to maintain the stability of a large-scale society, whether they be external controls applied by an elite of cultural engineers, or whether
they be some sort of systemic Malthusian social regulators, are
impersonal controls. They are as impersonal as the physical controls whose removal necessitated their existence, and they are
equally incapable of confirming man's significance.

This is a dilemma that was recognized but not resolved by Nor-
bert Wiener in 'The Human Use of Human Beings':

> The real danger [of machines à gouverner] is that such machines,
> though helpless by themselves, may be used by a human being
> or a block of human beings to increase their control over the
> rest of the human race or that political leaders may attempt to
> control their populations by means not of machines themselves
> but through political techniques as narrow and indifferent to
> human possibility as if they had, in fact, been conceived mech-
> anically. The great weakness of the machine - the weakness
> that saves us so far from being dominated by it - is that it
> cannot yet take into account the vast range of probability that
> characterizes the human situation. The dominance of the machine
> presupposes a society in the last stages of increasing entropy,
> where probability is negligible and where statistical differences
> among individuals are nil. Fortunately we have not yet reached
> such a stage.(20)

In spite of the fact that Wiener thought such a state undesirable,
all the energies of the discipline that he founded are directed towards overcoming the limitations of machines to deal with the
variety that characterizes the human situation in order to hasten
its arrival. In the meanwhile the cybernetician copes with this
human variety by denying it. The larger the scale of his problem,
the more variety he must deny. The scatter about the statistical
mean is ignored and the statistical differences among individuals
are made nil.

Even the most benign and democratic of rulers could not be concerned about the individual difficulties of hundreds of millions of
people without sufficient calories and vitamins, or even of 2 million
people without jobs. His or her compassion, by virtue of the very
scale of these problems, must become abstract. Abstract compassion, for the recipient, is unlikely to be either convincing or
consoling. The impossibility of conveying compassion to large

numbers of people is not a problem that can be overcome by improving the hardware of communications. Sentiments are not
easily transmitted across the boundaries that separate the different known worlds of the rulers and the ruled. Governments
in acquiescing in, or positively encouraging, the channelling of
vast resources into projects that increase mobility, are deliberately removing the single most effective constraint on the size
and intractability of such social problems.

Following a speech at Central Hall, Westminster (January 1972),
Paul Ehrlich was questioned about his association with the Club
of Rome, an organization that receives prominent and favourable
mention in the 'Ecologist's Blueprint for Survival'. The questioner
wanted to know the identities of this group of men who, it seemed,
were volunteering to manage us. Ehrlich denied that there was
anything sinister about the Club's intentions, but, when pressed
further, concluded that someone had to control things and he
would rather it were the Club of Rome than no one. The implicit
alternative was chaos.

But there is a second alternative to social control by environmental managers and cultural engineers. It is reasoned, persuasive argument. There is a world of difference between being persuaded to restrict the size of one's family because over-population
is a serious threat to the welfare of society, and being 'prevented'
by a cultural engineer 'from wanting so many children'. It is the
difference between a human world and a Brave New World. At
best, in his fantasies, the cultural engineer arrogates to himself
the powers of an Augustan deity to define the limits within which
man can be free to will. At worst he would be a puppet master in
control of a world that he completely determines. Argument that
seeks to alter attitudes, it might be argued, is simply another
method of control, no different in kind from the methods of the
cultural engineer. But this can be so only for the complete determinist whom we have already excluded from our debate. Certainly
we have conceded that attitudes and beliefs are social constraints,
but they are willed, self-imposed constraints. We cannot demonstrate that they are self-imposed; it is simply that if we deny
it, we deny man any meaning. He becomes a completely determined
product of his environment. Argument, when compared with cultural engineering, may seem a very weak method for altering
human behaviour. It is a method that grows weaker as societies
become more mobile. But it is the only method that acknowledges
the behaviour that it would alter as human.


CONCLUSION

The conventional wisdom of 'essential travellers' echoes that of
Lord Lauderdale; for them nothing could be more certain than
that every improvement in the machinery of transport and communications contributes to improve the conditions of people everywhere. Sir George Edwards in his presidential address to the

Royal Aeronautical Society in 1958 had the measure of those to
whom he was speaking: 'In my audience tonight are many important people who will have the responsibility of making decisions that
will decide whether in fact there are to be British supersonic
transports.'(21) The subsequent history of the Concorde project
shows that these people did not demur from the central thesis of
Sir George's address, that 'increasing speed in transport has
been the essential hand-maiden to increasing development all over
the world and that the only medium in which speed can continue
to increase is the air'. Concorde's commercial failure is attributed
by the adherents of this thesis to a combination of technical failure, political conspiracy, and bad luck. But their vision of the
nature of the development process to which Concorde was an intended hand-maiden appears unshaken.

The benefits of this development process have been most inequit-
ably shared. For every one of the world's citizens who has acquired a car, there are three others who, in the words of the
Director-General of the United Nations Food and Agricultural
Organization, are living 'below the margins of human existence'.
His observations to a conference in Rome in 1979 place the achievements and enthusiasms of the world's high-flying 'developers'
in a sobering context: 'Their struggle to survive, to collect a few
twigs to cook a handful of grain, is less dramatic in media terms
than the quarrels of car owners queuing for a few litres of petrol.
But it is of far greater consequence for the life of nations and
the peaceful future of our children.'(22)

Developments in transport technology that have made the subservient more accessible to the dominant, without at the same time
increasing the mobility of the subservient, have widened the
gulf between them. The devastation and demoralization of the indigenous societies of North America and Australia, the slave trade
and subsequent colonization in Africa, the residual genocide still
practised in Latin America, and the daily 8000-kilometre bombing
runs during the Vietnam War are but a few examples. In America,
the widespread distrust of those in authority, the breakdown in
communications between social classes, races and generations, and
the squalor of the lives of the immobile poor living in the ghettos
of one of the world's richest and most mobile societies, are but a
few of the reasons for doubting that yet more mobility for the most
mobile will make the world a better place.

The rudimentary machinery to which the original Luddites objected offered, as Lord Lauderdale proclaimed, the promise of greater
prosperity for all. What the Luddites were objecting to was the
betrayal of this promise - the concentration of more wealth in
fewer hands while large numbers of people were being forced
nearer to the margins of starvation. The promise offered by the
machinery of transport and communications has also been cruelly
betrayed. It promised not only greater prosperity, but greater
freedom - for all. It is a promise incapable of fulfilment. Almost
everywhere that new machinery of transport and communications
has loosened the constraint of distance, new machinery of social

control has developed to regulate the social and economic inter-course that has been made possible. This new machinery of social control, which must grow more powerful as mobility increases, is, by the very nature of the tasks that it is required to perform, undemocratic.

Those who are the most mobile are admonished not to pull the ladder up behind them. But the ladder is seen by the admonishers as a magic extension ladder that can be extended for ever. The formulators of official transport policies have had their heads figuratively and sometimes literally in the clouds. They have scarcely noticed that the lower rungs of the ladder, on which most of the world still perches, are being forced slowly into the mire.

## REFERENCES AND NOTES

1    Eltis, W. (1978), If We Want to Get Richer We Must Remove the Union Veto on Economic Growth, 'The Sunday Times', 17 September, p. 16.
2    Ibid.
3    'Hansard', 27 February 1812, House of Lords, col 966-72. This was Byron's maiden speech in the House of Lords.
4    Hughes, T. (1972), The World Shrinks at Supersonic Speed, 'Sunday Times', 16 January.
5    Beswick, Lord (1972), A Policy for British Aerospace, 'Flight International', 27 January, pp. 130-1.
6    'Hansard' (Lords), 22 January 1974, cols. 1283-4.
7    Ibid.
8    'Geographical Magazine', July 1972, p. 664.
9    Lighthill, Sir James (1971), Proceedings of the Eighth International Shock Tube Research Symposium, Imperial College London, 5-8, July 7/15.
10   Daher, M. Mikhael, MP for Akkar and Hermel, quoted in the Beirut newspaper 'Le Réveil', 12 May 1979.
11   Thoreau, H.D. (1854), 'Walden', (1960 ed.) Signet, New York.
12   Toffler, A. (1970), 'Future Shock', Bodley Head, New York.
13   Jones, W.T. (1969), 'The Medieval Mind' Harcourt, Brace & World, New York.
14   Gabor, D. (1971), Desirable and Undesirable Ends of Tech-nology, in 'Can We Survive Our Future? G.R. Urban (ed.), Bodley Head, London, pp. 196-206.
15   Seaborg, G.T. (1971), The Erehwon Machine: Possibilities for Reconciling Goals by Way of New Technology, in 'Energy, Economic Growth and the Environment', S.H. Shurr (ed), Resources for the Future, Johns Hopkins University Press.
6    Cited in Adams, J.G.U. (1973), Everything Under Control, 'Science for People', April/May.
7    Beer, S. (1971), The Liberty Machine, 'Futures', vol. 3, no. 4, an edited version of an address to the American Society for Cybernetics.

18   Goldsmith, E. (1971), Ecology, Controls and Short-term
     Expedients', in 'Can We Survive Our Future?' G.R. Urban
     (ed.), Bodley Head, London, pp. 207-24.
19   Beer (1971), op.cit.
20   Wiener, N. (1950), 'The Human Use of Human Beings',
     Anchor ed. 1954, Doubleday, New York.
21   Edwards, Sir George (1958), The Presidential Address,
     'Journal of the Royal Aeronautical Society', vol. 62.
22   Edward Saouma, (1979) quoted in 'The Observer', 15 July.

# APPENDICES

# I   ...AND HOW MUCH FOR YOUR GRANDMOTHER?*

The value of human life is a question that has always troubled
decision-makers. Directors of projects involving the loss of life
never have known just how to calculate the optimal level of such
loss. It is the great unsolved decision-making problem.

For most of recorded history, decision-makers who have faced
this problem have relied upon models of their projects that have
been subjective, intuitive, unsystematic, and imprecise. The one
notable exception was Jonathan Swift. In 1729 he published a
brief essay called A Modest Proposal in which he brought unprecedented mathematical precision to bear - he brought reason to
bear on the question. The value of human life was, he demonstrated, simply a function of supply and demand. Further, the optimal time to end it was that at which the selling price minus the
cost of production was a maximum. This time he computed to be
one year and the net profit on a plump yearling child he calculated very precisely at 40p (1729 prices).

Unfortunately he was ahead of his time. The combination of his
own modesty about the merits of his proposal and an unpropitious
academic climate resulted in the essay's complete neglect. It was
widely read as a satire and nothing more was heard of it in economics for over 200 years. (Although the method of evaluation employed by some twentiety-century researchers of subtracting consumption from production is remarkably similar to the Swift
method of deducting production costs from selling price, I can
find no acknowledgements to Swift in any of this century's literature on the subject; see for instance Dawson.(1))

Meanwhile decision-makers retained their time-confirmed habits
of subjectivism, intuition, and capricious imprecision. And History went its long, suboptimal way. The industrial revolution
gained momentum; roads, canals, and railways were built; dangerous rivers were bridged and mines dug; the West was won and
empires subjected; wars were fought and armies sacrificed - all
without a workable procedure for determining whether or not the
sacrifice of human life involved was even approximately optimal.(2)

And so the world might have continued but for the advent of
two significant factors: democracy and the computer. The infan-
trymen and peasants of past ages who were sacrificed in such
numbers had no voice in the councils that decided their fate. But
today their modern counterparts want the decision-makers who
declare wars on their behalf to justify the sacrifices they ask of

*First published in 'Environment and Planning A', 1974, vol. 6.

them. The masses are demanding to be convinced that the benefits do in fact outweigh the sacrifices they are called upon to
make.

The growing scale of modern warfare has complicated the
decision-maker's problems enormously, but computers have come
to his aid. With machines that can perform millions of calculations
in a fraction of a second, the scale of a military project need no
longer affect the precision with which its benefit-cost ratio may
be calculated. Not only wars but a great variety of civilian projects involve sacrifices measurable in megadeaths. Road schemes,
airports, nuclear-power generators, and a great range of polluting industrial activities all confront the environmental planner
with the task of reconciling the benefits expected to result from
the projects with the cost of death. Such problems can now be
brought within the compass of optimizing decision theory - provided one hitherto insurmountable obstacle can be overcome: the
calculus of optimization requires that all costs and benefits be
measured in the same units. Any units will do but convention suggests money. Thus, before the apparatus of decision theory can
be brought to bear on such projects, it is imperative that we be
able to place a cash value on human life.

In recent years there has been an impressive growth in the
number of academic papers devoted to this most intractable of
valuation problems,(3) but they have all been bedevilled by the
same fundamental misconception. They have as a result, and as
Mishan (4) has scathingly demonstrated, been comic failures. They
have all foundered on the difficulty that, in order to be logically
consistent, they require that a man be prepared to place a cash
value on his own life, and this, common sense tells us, is absurd.

*A theoretical breakthrough*

In a recent paper, entitled Evaluation of Life and Limb: a Theoretical Approach, Mishan sweeps away all the confusion that has
built up around this misconception and then lays the logical
foundation necessary for a rational discussion of the issues involved. The paper is, I think it is fair to say, the definitive work
on a very important subject, and as such has not received the
public attention it deserves. My purpose therefore in the remainder of this paper is, in a spirit of frank admiration of Mishan's
theoretical achievement, to suggest a number of practical implications of his approach and then to elaborate some additional
theoretical conclusions that might be drawn from it.

But first a brief summary of the most important arguments in
his paper. He begins with a broad survey of the literature on the
topic and dismisses all of it as 'inconsistent with the basic rationale
of the economic calculus used in cost-benefit analysis', namely
the potential Pareto improvement criterion. This he demonstrates
simply and effectively with the help of the following formula:

$$\sum_{j=1}^{n} V_j > 0 \qquad\qquad [1]$$

If $V_j$ is the maximum sum that an individual $j$ is prepared to pay for the benefit he will derive from a given project (or, if he will be adversely affected by the project, the minimum sum that he will accept as fair compensation) then the summation of these values for all individuals (with compensation values prefixed by a minus sign) yields the net value of the project. If the sum is greater than zero then the project will produce a potential Pareto improvement. Clearly, if a project were to involve the loss of a specific life, the value of $V_j$ for the individual specified would be minus infinity, and no conceivable collection of benefits could mak $\Sigma V_j$ positive. The project would therefore be rejected as failing to meet the Pareto criterion.

This formula, it might be mentioned in passing, applied to military decision-making problems, provides a very useful scientific criterion for distinguishing the civilized from the barbaric races of the world. It is well known that military commanders belonging to barbaric races, such as the Japanese, are prepared to sacrifice specific men in battle. But civilized commanders, imbued with the ideals of Pareto optimality, deny themselves the advantage of sacrificing specific men for the cause. Although it might be certain that a given military exercise will cost lives, every individual involved must be given a statistical possibility of surviving. Otherwise the exercise could not produce a Pareto improvement.

This military example can serve to illustrate the most important concept in Mishan's paper. While decision-makers in Western democracies are forbidden, by a well established ethical tradition, to undertake projects that will result in certain death for individuals whose identities can be known in advance, they can and do undertake projects that involve the certain loss of life. For example, in this country on an average day approximately twenty people are killed in road accidents. At the beginning of the average day the magnitude of the number of fatalities is known but the individuals comprising it are not. Similar examples could be cited for a great number of other risky ventures that have been studied by statisticians, such as mining and bridge building.

It is Mishan's discovery of this critical distinction, between the certainty of death in the aggregate and certainty in the individual case, that has removed the rock on which all previous studies in the field have foundered. It was both a logical and an emotional rock. It was a logical obstruction in the sense that a failure to grasp this distinction led to conclusions that were logically inconsistent with Pareto optimality. Mishan underlines the revolutionary importance of his discovery:

> The basic concept introduced in this paper is not simply an alternative to, or an auxiliary to, any existing methods that have been proposed for measuring the loss or saving of life. It is the only economically justifiable concept. And this assertion does not rest on any novel ethical premise. It follows as a matter of consistency in the application of the Pareto principle in cost-benefit calculations.

It was an emotional obstruction in the sense that the work of all

the pioneers in the field, including Swift, offended the sensibil-
ities of people who felt that it was callous to talk cold-bloodedly
about the cash value of specific people. The removal of the need
to talk about specific lives has thus removed a serious impediment
to the rational discussion of sacrifice; although our ethical tradition strongly disapproves of the taking of life in the particular,
there is nothing in this tradition that particularly opposes the
taking of life in the abstract, so long as the price is right.

Let us now look more closely at Mishan's achievement of logical
consistency within a cost-benefit framework. The argument, he
says, 'which has it that a man who accepts £100 000 for an assign-
ment offering him a four-to-one chance of survival will agree to
go to certain death for £500 000, is implausible to say the least'.
The vital point here is that, while rational people refuse to accept
any amount of cash compensation for certain death, they will
accept cash compensation for undertaking certain risks.

There is a great wealth of subtlety and technical sophistication
in the argument that Mishan builds on this point. For the sake of
brevity I will set out what I take to be its essence. $V_j$ is replaced
by $r_j$ and a new formula is presented:

$$\sum_{j=1}^{n} r_j,$$

where $r_j$ is the amount of compensation that an individual, $j$, will
accept for the undertaking of a specified risk. In general this
will be prefixed by a minus sign. But on occasions where $j$ stands
to benefit, perhaps from a project that threatens someone who
might remember him in his will, $r_j$ will be positive. The compensation values for all parties involved are added together and set
against the benefits of the project and, if the final total is positive,
then the project will produce a Pareto improvement.

*The risk-compensation function*
Mishan suggests that $r_j$ will be some nonlinear function of risk. At
this point I would like to present my own personal risk-compensation function. This may strike the reader as an unwarranted departure from the dispassionate detachment with which the discussion has been conducted so far. However, as Mishan observes,
very little is known about the cash values that people subjectively
attach to risks, and the economist will have to resort to asking
them in order to find out. Thus, having given considerable
thought to the way in which I rationally value risks to my own life,
I offer the function described below as raw data with which to
begin the filling of the empirical void.

My function I have discovered is really quite simple; it has the
following form:

$$r = a\,(x-p)^{-b},$$

where $r$ is compensation in pounds sterling, $x$ is the size of the population among whom the risk of one death is distributed (that is the reciprocal of the risk of death to any individual in the population), $p$ is the size of the population below which no amount of compensation would be a sufficient inducement to me to undertake the risk (that is the reciprocal of the maximum risk I will willingly subject myself to), and $a$ and $b$ are additional parameters determined empirically.

The specific values of the parameters in my function I obtained empirically by consulting myself. They are illustrated by Figure I.1 It can be seen by reference to this graph that I would be prepared to play Russian roulette with a 1001 cylinder revolver for £1 000 000 but will not willingly subject myself to a risk greater than 1 to 1000 for any realistic amount of money. At the other extreme I find a risk of 1 to 100 000 too small to be meaningful

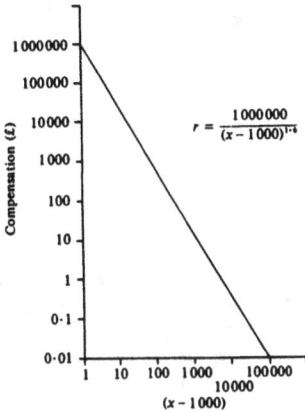

Figure I.1

and would demand a token 1p before subjecting myself to it. Risks even smaller than this I will subject myself to for nothing. The value of 1·6 for exponent $b$ is greater than 1, indicating that as risk is increased, and death stares me ever more menacingly in the face, the amount of compensation I will demand increases at an even faster rate. Although Mishan does not himself attempt to define an explicit risk-compensation function, my function presented here is consistent with his speculations about the price of risk discussed above.

One can easily ascertain, by conducting a straw poll of one's associates, that the values of $a$ and $p$ vary enormously. (The result of my poll was complicated by a very large percentage of 'don't knows'. This complication is discussed below.) $a$ seems to be a direct function of wealth; the distribution of $p$ appears to be U-shaped, with pessimists assigning very large values to $p$ and optimists very low values, with a very few moderate people, such as myself, choosing an intermediate value such as 1000. But,

assuming that the basic form of the function is typical of most
people's reaction to risk, and assuming $b$ to be always greater
than 1, some very useful conclusions can be drawn that will be
valid whatever the specific values of the parameters.

*Economics of scale and economies of ignorance*
An astute manipulation of their statistics will yield planners hitherto undreamed of economies of scale and ignorance. Looking first
at the economies of scale it can easily be seen that the total compensation that must be paid for a given number of expected
fatalities $\left( \sum_{j=1}^{n} r_j \right)$ will decrease as $n$ is increased. In fact, if the

planner can manage to spread the expected deaths over a
population sufficiently large to reduce the individual risks to
negligibility, no compensation will have to be paid at all. We can
anticipate then that the ever-increasing scale of economic and
political activity in the world and the increasingly nonselective
nature of both military and civilian modes of killing will steadily
reduce the specific probabilities of death associated with specific
projects and drive the cost of life steadily down.
 For those impatient with the slow rate at which the natural
growth in the scale of society is reducing the cost of life, there
remain to be exploited a large number of economies of ignorance.
The more precisely the population at risk from a given project
can be specified, the greater the compensation the specified population will demand. For example, if it were known that a lead-
smelting operation would poison only about twenty people in this
country, the average level of risk would be only about 1 in 3 million,
a level that most people in this country would consider negligible.
But if it were known that this group of twenty would all be found
among the 100 furnace tenders at the smelter, some among the 100
might place inordinately high compensation values on the risk to
which they were subjected and the smelter might have to be closed.
 People such as these might on occasion even place infinite compensation values on only moderate risks and thus disqualify on
Pareto optimality grounds projects that would otherwise be of
enormous social benefit. Short of lying or withholding information
about risks there has apparently been no way of dealing with such
people. If the Pareto principle were to be strictly adhered to,
then fear would remain, it seems, an insurmountable barrier to
progress. But, there are other, perfectly honest, ways of reducing the costs of risk compensation without actually reducing the
number of deaths. Consider an illustration suggested by Mishan,
a hypothetical case in which it were known that there was a very
high probability that within a given period of time a sonic boom
from a Concorde would trigger fatal heart attacks in certain medical cases living beneath the flight path. Let us further hypothe-
size that for some flight stages no flight path could be plotted that
would not have at least one of these cases. Pareto optimality considerations would demand the sacrifice of all the enormous benefits

of supersonic travel for the saving of only one or two lives. But, if a flight path randomizer could be developed that would produce constrained random deviations from the flight path and reduce almost certain specific death to an acceptable level of risk, the project could be allowed to proceed.

Great strides have recently been made in the application of this principle to the evaluation of nuclear energy projects. Although the radiation associated with their operation and waste entail the certainty of large numbers of premature deaths, the stochastic nature of the way in which radiation selects its victims, makes it impossible to be specific about the identities of the people who will die. This greatly reduces the cost of death associated with such projects. But even in this field the economics of ignorance are as yet very imprecisely quantified. They remain, at the moment, exciting theoretical possibilities for the future rather than quantities of immediate operational value. Regrettably, I must conclude with an examination of some of the mundane operational problems that still stand in the way of a more optimal future.

*The research frontier*
Although the conceptual foundations laid by Mishan for the valuation of the loss of life are very strong, before such valuations become a practical possibility for general use in cost-benefit studies some rather intractable measurement problems must be overcome. I have reserved the discussion of the difficulties for the end, but the problems raised must not be taken as an excuse for defeatism. Rather they should be viewed by all economists as a challenge that defines an important research frontier. I repeat the caveat of Mishan: 'The problem of measurement must not be allowed to obscure the validity of the concept.'

It is generally argued that the valuation of life that is to be fed into cost-benefit studies and used for policy decisions must be that of society as a whole and not just the valuation of a few experts. But eliciting such a value from society is no easy matter. First, not a single figure, but a finely graded tariff is required. Society values old people, who are going to die in the near future anyway, less highly than people who are in their middle years. People in their middle years are also valued more highly than newborn infants because society has made a larger economic and emotional investment in them. The problems of questionnaire design are formidable. Because, as we have noted above, lives in the abstract tend to be valued less than specific lives, the questionnaire must be very specific: 'How much for your best friend, neighbour, greengrocer . . . spouse, child, uncle . . . and grandmother?' are questions of the sort that will have to be answered if reliable estimates are to be obtained. Even then problems will remain. For example, cultural weightings will have to be devised to compensate for the fact that some cultures value the extended family more highly than others; key words will have to be tested for emotional bias; time discount rates will have to be estimated for lingering death situations; answers will have to be

standardized for age, sex, and education; equity problems will
have to be resolved. . . . The list is a long one.

However, even after all the work on the problems of questionnaire design and administration is completed, serious problems
of questionnaire interpretation will remain. The difficulty with
asking people questions such as 'How much for your grandmother?'
is that they do not generally give honest answers. If they think
that reporting a high value might influence the government to
spend more money on the welfare of their grandmother, and they
want their grandmother better cared for, they will answer accordingly. But if they think such an answer might result in higher
taxes for the possessors of grandmothers then they will report
very low values. However, this may not be the insuperable problem it seems at first glance. A positivist economist will not generally know in advance what use will be made of the answers to
the questionnaire since his job is only to present the alternatives
and advise on optimal strategies for somebody else's value judgments. So, making a virtue of necessity, a new school of thought
has arisen to argue that we should also make a virtue of ignorance.(5) If we genuinely cannot tell people what use will be made of
their answers then they will not be able to cheat. So long as we
remain ignorant of even the probabilities of future policy decisions
there is no possibility of self-interested bias creeping into the
answers.

This solution to the cheating problem does, however, add to
another problem, the last that I wish to discuss. This is the problem of what to do with the 'don't knows', a category that plagues
the lives of all practising opinion pollsters. If a man with a fixed
income has a number of hypothetical questions about expenditure
to answer simultaneously (as is commonly the case in reality) then
things get computationally somewhat complex. But even when very
simple valuation questions were put, well over 95 per cent of
those questioned in my straw poll fell into the 'don't know' category. This very widespread inability to perform the calculations
necessary to convert lives into money is a very worrying reflection of the unworldly nature of the education system in this
country. But it also raises an interesting epistemological question:
namely, where does the question 'What is the value of a human
life?' come from? Mishan suggests that it emerges from political
debate; the people want to know and ask the economist, who is
society's acknowledged expert on the cash value of things. But
because the answer is subjective and locked up inside people's
heads the only way the economist can get at it is by asking the
people. The problem Mishan acknowledges, is 'somewhat circular'
This circularity is a serious problem, but, fortunately, not an
insurmountable one. Philosophers who have had long experience
with such subtle issues are reassuring. As one has noted
(Bertrand Ryle), 'If a dog turns in circles quickly enough he
usually succeeds in catching his own tail'.(6)

The difficulties, then, are many but the potential rewards are
great. Cost-benefit analysts on the research frontier must become

multidisciplinary men embodying the skills of not only the economist but the philosopher, sociologist, psychologist, doctor, and
natural scientist as well. We must not relent. Mishan reminds us
of the importance of the task upon which we are embarked; 'As
cost-benefit studies grow in popularity, it is increasingly important to make proper allowance for losses or gains arising from
changes in the incidence of death, disablement, or disease caused
by the operation of new projects or developments.' We must not
be deterred by the softhearted among us who prefer not to think
of death and disablement in terms of money. Rationality and
efficiency demand that we reduce everything to cash. If we refuse,
we throw away all the inestimable benefits of the cost-benefit
calculus.

## REFERENCES AND NOTES

1   Dawson, R.F.F. (1967), 'Costs of Accidents', Road Research
    Laboratory Report, LR 79.
2   Some would cite the calculations of those engaged in the
    Atlantic slave trade as a counter example to this historical
    generalization. But the criticism misses the point that we are
    here concerned only with the rational valuation of human life.
3   See, for example, the useful bibliographies accompanying the
    work of Dawson (1967), op.cit., Mishan, note 4 below, Fromm,
    G. (1965), Civil Aviation Expenditures, and Klarman, H.
    (1965), Syphilis Control Programs. The last two are found in
    'Measuring Benefits of Government Expenditure', R. Dorfman
    (ed.), Brookings Institution, Washington DC.
4   Mishan, E. (1971), Evaluation of Life and Limb: a Theoretical
    Approach, 'Journal of Political Economy', 79, pp. 687-705;
    also printed in 'Cost Benefit Analysis', Unwin University
    Books, London.
5   This is a theme pursued by Bohm, P. (1971), An Approach to
    the Problem of Estimating Demand for Public Goods, in 'The
    Economics of Environment', P. Bohm and A. Kneese (eds),
    Macmillan, London. In 'A Theory of Justice' (Clarendon Press,
    Oxford, 1972), philosopher John Rawls suggests that the
    construction of a just social framework can only be achieved
    if the builders are kept in ignorance about the consequences
    of the framework they are building. The recent rediscovery
    of the value of ignorance appears likely to open up a large
    number of exciting prospects in the social sciences.
6   Ryle, Bertrand, 'Tautologies I have Known', Reason Press,
    Oxford, out of print, date unknown.

# II  WHAT NOISE ANNOYS?*

Noise is in the eye of the beholder. Consider the evidence displayed in Figure II.1. It comes from a study done in 1971 of the reactions to noise of 693 people living in Paris. The study is described as one of the most successful attempts ever made to correlate noise levels with noise nuisance, in a book entitled 'Road Traffic Noise'.(1)

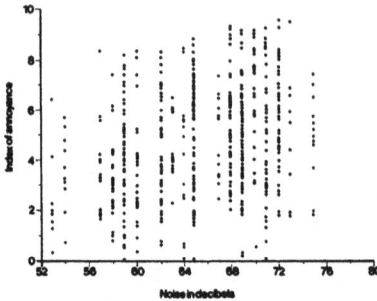

Figure II.1

There are obviously a number of ways in which the dots might be connected. Figure II.2, a vole clutching its ears while standing in the rain, illustrates one method.

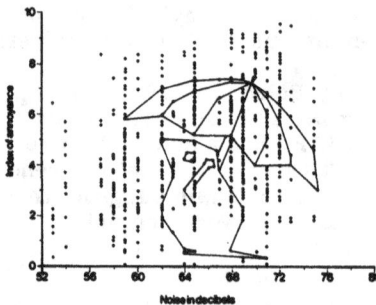

Figure II.2

Figure II.3 three jagged lines and one straight line standing in the rain, illustrates the method of statistical science.

The choice of method has become of some importance. Cheap,

*First published in 'Vole', 1, 1977.

miniaturized sound level meters are widely available, and no
public inquiry into a road or an airport is now deemed complete
without a large supply of measurements produced by them. At
the Archway Road Inquiry in north London, for example, the
Department of Transport presented the following information in
support of its proposal: in the year 2000 the Department's scheme
would produce 0.5 decibels less noise outside Goldsmith's Court
than the alternative scheme proposed by the Borough of Haringey,
1 decibel less in Priory Gardens, and 1 to 2 decibels less on the
Archway Road between Muswell Hill Road and Shepherd's Hill.
What all this means depends on how you join up the dots.

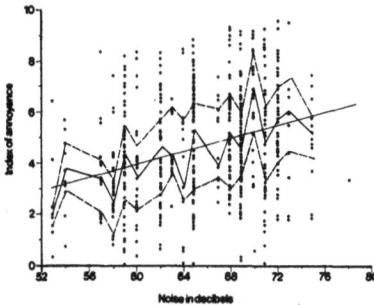

Figure II.3

The method illustrated by Figure II.2 has, at the moment, no
known adherents within the Department of Transport, so let us
dwell briefly on the method of statistical science. The outer jagged lines of Figure II.3 embrace half of the 693 dots; the distance between them is referred to as the interquartile range. The
central jagged line runs through the median annoyance score for
any given noise level. The straight line, called the 'line of best
fit', is the line you use if you want to predict the way in which
the level of annoyance would change if noise were to increase or
decrease. Thus if you want to predict the way in which the level
of annoyance of an average person living in, say, Priory Gardens,
might change if his noise exposure were to be increased from 60
to 61 decibels, you simply consult the line to discover that it
would increase by about 0.15 units of annoyance. The method
used to fit the line assumes that these units of annoyance have a
constant size and can be added, subtracted, multiplied and divided.

The fact that no one has ever seen a unit of annoyance is a
problem. They have proved more elusive than quarks, and it is
admitted that, technically speaking, one is not really entitled to
assume that they exist at all. But entitled or no, it is argued in
'Road Traffic Noise' that sheer necessity requires them:

To make practical predictions, the gap between ordinal and
cardinal numbers has to be jumped. And from sheer necessity
this requirement overrides that of certainty for all assumptions
upon which our procedures rest.

Sheer necessity is the mother of sheer invention. It remains, however, for it to bring forth a meaning for the data in Figure II.1 -
or for the numbers presented to public inquiries by the Department of Transport. A lot of earnest scientific endeavour has produced a bewildering lack of evidence that sound, below the level
at which it causes pain or physical damage, correlates convincingly with anything at all. The biggest book I have been able to
find on the subject, 'The Effects of Noise on Man' by K.D.
Kryter,(2) after 586 pages puts it this way:
> A possible teaching of much of the data presented in this book
> is that, other than as a damaging agent to the ear and as a
> masker of auditory information, noise will not directly harm
> people or interfere with psycho-motor performance. Man should
> be able, according to this concept, to adapt physiologically to
> his noise environment, with only transitory interference effects
> of physiological and mental and motor behavioural activities dur-
> ing this period of adaptation. This concept, or its converse, is
> difficult to substantiate by scientific research and must be
> recognised as being hypothical [sic] at this time.

The fact that the tale thus far signifies nothing, far from discour-
aging the scientists, is a spur. They remain convinced that lurking somewhere in their data there must be a relationship which,
once discovered, will permit the 'practical prediction' of the way
people will respond to changes in sound. This conviction is the
basis upon which large amounts of public money are procured to
pay for the search. Meanwhile, present lack of success in explaining what their sound measurements mean has not deterred the
scientists from offering the world large numbers of them. Surely
they must mean something.

Instructions for the measurements of traffic noise are contained
in a book produced by the Department of the Environment entitled
'Calculation of Road Traffic Noise'.(3) They are lengthy, detailed
precise, and expensive to implement. Since noise is widely acknowledged to be a major nuisance and irritant, and since the
Department has gone to so much trouble and expense devising
ways to measure it very precisely, and has gone to the further
trouble and expense of creating a body of noise compensation
legislation based upon these measurements, one could be forgiven
for assuming that the Department's noise measurements give a
good indication of the nuisance and irritation caused by noise.
They do not.

The belief that they do, and incorporation of the belief in compensation legislation and road scheme appraisal methods can be
attributed, in part at least, to the conclusion of the Department's
Urban Motorways Project Team(4) that the measurements 'relate
fairly closely to what we know about public attitudes'. What they
knew about public attitudes to traffic noise rested, the team
said, on a 1968 study by the Building Research Station. The BRS
study concluded as follows: 'Individual dissatisfaction scores cor-
related poorly with physical measures. This finding is believed to
be the result of wide individual differences in susceptibility to

and experience of noise, as well as in patterns of living likely to
be disturbed by noise. Attempts to allow for these factors were
unsuccessful.'(5)
Concorde has been the subject of more sound-measuring exercises
than any plane in aviation history. That the exercises are pointless can be simply demonstrated. Collect a group of people consisting half of ardent supporters of Concorde (such as John Ran-
kin, MP, who insists that the plane is 'one of the quietest pro-
ductions aviation has ever created') and half of vehement opponents (such as Richard Wiggs of the Anti-Concorde Project who
insists that it is one of the noisiest). Equip them with a sound-
meter and stand them all together in Green Man Lane in Hatton
at the eastern end of Heathrow's southern runway. Get them there
in time to witness a sample of subsonic jets taking off before Concorde. Then wait for Concorde.
     As soon as it passes overhead a remarkable thing happens. Half
the group clutches at its ears and collapses in agony. The other
half starts leaping up and down with a manic gleam in its eye.
The figure on the sound meter is irrelevant. What is being reacted
to is not the intensity of the sound but its message. To half of
the group it is the roar of progress. Its note is the uplifting one
of freedom and mobility, and science in the service of man. To
the other half it is the thunder of technology gone mindless, and
its note is the ominous one of civilization in retreat, before the
advancing multinational, off-shore jet-set. The frequencies and
acoustic energies of both sounds are the same.
     The Department of the Environment is taking the matter in hand.
It has appointed a Noise Advisory Council to advise it, and is
formulating a framework within which noise control policy can be
formulated. In a foreword to the Noise Advisory Council's report
'Noise in the Next Ten Years' (6) Denis Howell, the Department's
Minister of State, sets out the government's view of the noise
problem:
>     In the long run the most effective element in noise control policy
>     is to reduce noise at the source itself. The increase of noise
>     nuisance is largely a by-product of our growing dependence on
>     machinery, and industry must be encouraged to develop quieter
>     vehicles and engines and to provide proper insulation for noisy
>     machinery. To this end the Government is spending large sums
>     on research.

The Council expresses no concern about our growing dependence
on machinery. It does, however, betray concern that insufficient
numbers of people share its view that we should grow more dependent on machinery quietly. It recommends that 'Public opinion
must be continually stimulated to recognise the importance of
noise and the possibilities for its alleviation'. The sound of road
traffic is the sound of machinery that every year kills and injures
over 300 000 people in Britain. It is also the sound of people abandoning their local shops for hypermarkets, of irreplaceable oil
being burned, and of invisible lead and carbon monoxide being
spewed into the air. The sound of air traffic is the sound of the

machinery of the tourist industry, and of the Kissingers and Yamanis and Frosts going about their daily business of running the world. They are the sounds of anomie on the march as every day more strangers pass one by. Noise is unwanted sound. We all have to some degree the ability not to hear things we do not want to know. We now have the help of the government. The government's noise control policy is to cut out the tongue of every messenger who brings bad news.

## REFERENCES AND NOTES

1   Alexandre, A. et al. (1975), 'Road Traffic Noise', Applied Science Publishers.
2   Kryter, K.D. (1970), 'The Effects of Noise on Man', Academic Press.
3   Department of the Environment (1975), 'Calculation of Road Traffic Noise', HMSO.
4   Department of the Environment (1973), 'Report of the Urban Motorways Project Team to the Urban Motorways Committee', HMSO.
5   Langdon, F.J., and Griffiths, I.D. (1968), Subjective Response to Road Traffic Noise, 'Journal of Sound and Vibration', vol. 8, 1, pp. 16-32.
6   Noise Advisory Council (1974), 'Noise in the Next Ten Years', HMSO.

# III  SEAT BELTS

More than twenty countries have made the wearing of car seat belts compulsory. Had the general election not intervened in May 1979, the Road Traffic (Seat Belts) Bill which gained a Second Reading in Parliament on 22 March with a majority of 244 to 147, would almost certainly have become law, making their wearing compulsory in Britain as well.

The debate about seat belts has been primarily between those who argue that the benefit in lives saved justifies compulsion, and defenders of civil liberties who are opposed to the state intervening to protect people from themselves. But both sides of the parliamentary argument, with the notable exception of Enoch Powell, have been agreed that making the wearing of seat belts compulsory would save a significant number of lives.

The most authoritative and comprehensive survey of the benefits of seat belts published to date in Britain is a report from the Transport and Road Research Laboratory entitled 'The Protection Afforded by Seat Belts' (SR449, 1979). This report concludes that making the wearing of seat belts compulsory in Britain would save about 630 deaths and 12 900 serious injuries every year. In the Second Reading debate in Parliament these numbers were subjected to generous rhetorical rounding. In introducing the Bill the Secretary of State for Transport made the following claim:

> On the best available evidence of accidents in this country - evidence which has not been seriously contested - compulsion could save up to 1,000 lives and 10,000 injuries a year. . . . Every day this bill is delayed, two or three persons die who need not die and two or three families are bereaved.

Let us, for the moment, accept the claim of a possible 1000 lives a year which would be saved by making the wearing of seat belts compulsory. In Britain in 1977 there were an estimated 210 000 million passenger kilometres travelled by car. If we make the generous assumption that one-third of these kilometres were travelled by voluntarily belted motorists, this leaves 140 000 million kilometres travelled by un-belted motorists. Thus a measure that would reduce by 1000 the number of lives lost in all this travelling would reduce the risk of fatal accident per kilometre travelled by one 140-millionth. Many trips are considerably shorter than one kilometre, so the additional risk incurred by making such trips unbelted would be considerably less than one in 140 million. It would have been the effect of the Seat Belts Bill to make the taking of such risks a criminal offence.

That a person travelling at speed inside a hard metal shell will

stand a better chance of surviving a crash if he is restrained
from rattling about inside the shell is both intuitively obvious
and supported by an impressive volume of empirical evidence.
But the evidence for the efficacy of seat belts that is presented
in the TRRL report relates, with an important exception which
will be discussed below, to the enhanced chances of surviving
crashes in cases where seat belts are worn voluntarily. It ignores
the possibility of an increase in risky driving associated with the
additional sense of security provided by seat belts, and it ignores
the even more difficult problem of assessing the effects of compelling people to do something that they would not do voluntarily.
Since those who do not now wear seat belts have shown themselves
resistant to a very considerable campaign of persuasion, it might
be presumed that they would, at best, comply with the letter of a
new seat belt law rather than with its safety spirit. But this presumption, and the contrary presumption of the advocates of compulsion, can only be tested by an actual experiment. The experi-
mental evidence presented in the TRRL report all comes from
Australia.

Australia was the first country in the world to make the wearing
of seat belts compulsory and is therefore the country in which the
consequences of the experiment have been monitored for the longest period.

The State of Victoria made their wearing compulsory on 22
December 1970. New South Wales on 1 November 1971 and all the
rest of the country by 1 January 1972. A report by the Australian
House of Representatives Standing Committee on Road Safety,
'Passenger Motor Vehicle Safety' (Parliamentary Paper 156, May
1976), was well pleased with the result. It concluded that the
value of the seat belt legislation has been 'proved' and cited percentage reductions in injuries and in lives both similar to those
reported in the TRRL report.

Figure III.1 displays the most important of the Australian evidence that has been offered as 'proof' of the claims for the seat
belt legislation. It is proof not so much of the efficacy of the legislation as of the power of statistics to befuddle people.

The savings of life and limb that are claimed are not reductions
in absolute numbers but are reductions below death and injury
levels that the statisticians have hypothesized would have occurred
had the seat belt legislations not been passed. These hypothetical
levels were calculated by fitting a straight line to an arbitrarily
chosen number of data points for previous years and calling this
a 'trend'. The importance of the set of years chosen for the calculation of the trend is demonstrated by lines *a* and *b* applied to
the data for New South Wales. Line *a* is fitted to the period 1961
to 1970 and line *b* to the period 1965 to 1971. A much greater
variety of possible 'trends' can be found lurking in the British
data for the same period. Lines *c* and *d* illustrate the sensitivity
of the method to the assumption that the trend line is linear. If
the line is allowed to bend slightly to conform more closely to the
slight curve in the data for the period from 1961 to 1971, it can,

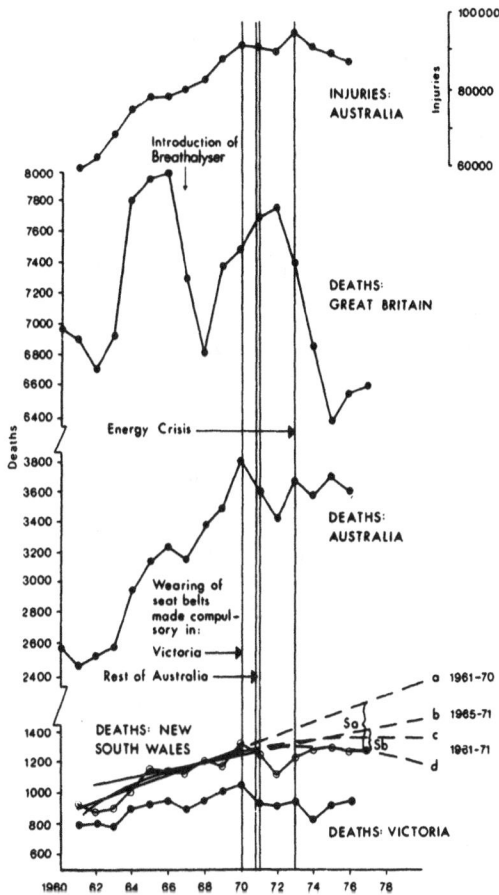

Figure III.1

when projected, be made to flatten out or even turn down. There are no laws of nature, either physical or human, that can be appealed to for guidance in deciding what shape of line ought to be fitted to what set of years. The choice depends entirely on the preconceptions of the statistician about the nature of the process he is modelling. The trend line, once fitted is extrapolated into the period after the legislation came into effect, trends apparently being things that go on for ever if not interfered with. The distance between this hypothesized level and the actual death level is measured and expressed as a percentage of the hypothesized level. This difference is called a 'saving' and is attributed to the seat belt legislation. A comparison of the 'saving' measured against line *a* (Sa) with that measured against line *b* (Sb) shows that the saving calculated by this method can vary enormously depending

on the arbitrary way in which the trend has been calculated and
extrapolated. The fact that there were actually more deaths and
injuries in some years after the legislation came into effect than
before is nowhere even mentioned in the exhortatory reports describing the savings resulting from compulsory seat belt wearing.
The reports also fail to call attention to the fact that the level to
which the Australian death rate was 'reduced' by 1976 was more
than twice as high as the death rate in Britain in the same year:
26 per 100 000 in Australia. 12 per 100 000 in Britain.

Evidence in the Australian report yields some prima facie support, albeit unwitting, for the hypothesis that compulsory safety
measures tend to be compensated for by less cautious driving.
Paragraph 430 of the report observes:

It should be noted that fatalities and injuries of other road users
(pedestrians, cyclists and motorcyclists) did not show a cor-
responding decline either in Victoria or in the rest of Australia.
In fact, during 1974 other road user deaths increased by 18 per
cent from the levels predicted and injuries for this group increased by 33 per cent. This strongly confirms the contention
that vehicle occupants were being affected by a measure not
operative so far as other road users are concerned.

Thus, while all reductions below their trend lines are claimed
as benefits attributable to compulsory seat belt wearing all increases above the trend are disclaimed as having nothing at all
to do with the legislation: the seat belt measure, so far as other
road users are concerned, is declared 'non-operative'.

Those contemplating making the wearing of seat belts compulsory in Britain should indeed be interested in the hypothetical
question of what death levels in Australia might have been if the
legislation had not been passed. The graph of road accident
deaths for Britain, superimposed on the Australian data in Figure
III.1 for the same period, shows that in the same period between
1971 (the year in which the wearing of seat belts was not made
compulsory in Britain) and 1976 there was a reduction in the
death rate of 15 per cent. In Australia in the same period the
reduction was 0.2 per cent. Clearly all other accident-related circumstances were not the same in the two countries over this period.
But in both countries drivers were subjected to propaganda
about the petrol-wasting consequences of heavy-footed driving;
in Britain and in parts of Australia speed limits were lowered;
and throughout both a number of other road safety campaigns,
especially related to drinking and driving, were aimed at reducing the death toll.

Some complex set of factors other than the compulsory wearing
of seat belts reduced the road death toll in Britain over a six-
year period by a percentage 75 times greater than in Australia
in the same period. Whatever these factors may have been, the
claims that have been made for the effectiveness of seat belt legislation are based on the assumption that none of them has been
operating in Australia since the beginning of 1972.

The number of people killed in traffic accidents in Britain rose steadily, with only minor fluctuations in the graph, from 4513 in 1948 to 7985 in 1966. The sharp drop in 1967 and 1968 was associated with the introduction of the breathalyser. The effect of the breathalyser began to wear off very rapidly as traffic continued to increase and people came to appreciate that the discretion exercised by the police in the use of the breathalyser was such that there was an extremely small chance of getting caught while over the limit. The 1966 peak had almost been regained when another sharp drop occurred. The most likely causes of this second drop are the temporary decrease in traffic, the temporary reduction in speed limits, and the increased general awareness of the economic benefits of gentle driving that followed the 1973 energy crisis. But whatever the causes of the post-1973 drop, they are now wearing off.

The effect of the legislation making the wearing of crash helmets compulsory for motor-cyclists provides the most relevant British evidence bearing on the question of the efficacy of compulsory self-protection laws. The law came into effect on 1 June 1973. In 1974, the first full year in which the measure had been in force, fatal accidents to motor cycle riders were 9.6 per cent higher than in 1973, an increase from 641 to 703. Deaths for all other categories of road user fell in the same year by 7 per cent. Part of this difference is explained by the fact that motor cycle traffic was the only class of traffic to increase in 1974, but only by 6.1 per cent. Given the quality of the data and the great range of factors that could have had a bearing on motor cycle accident rates in the period since 1973 it is impossible to isolate with confidence the effect of the crash helmet law. But the evidence, such as it is, can scarcely be said to constitute support for the argument that it has saved lives.

The very most that can be claimed for the Australian seat belt legislation on the basis of the available accident statistics is that it coincided with a stabilisation of the road death toll at a level that is one of the highest in the world. But the 'success' claimed for the legislation has provoked a rush of emulators in legislatures around the world.

The curious readiness of most British legislators to believe claims based on such flimsy statistical evidence is perhaps a sign of an uneasy parliamentary conscience. The most remarkable aspect of the claims is their acceptance of an ever-rising toll of road deaths as the norm against which success ought to be measured. Traffic accidents are caused by traffic. The underlying cause of the growth in traffic deaths since the war has been the increase in traffic. This increase has been fostered, and is still being fostered, by a wide range of government initiatives, the most prominent being the lavish assistance provided for the car industry and the even more lavish expenditure on the road programme; it has been the policy of both Conservative and Labour governments, almost completely unopposed in Parliament, to promote the principal cause of traffic accidents.

# INDEX

For Product Safety Concerns and Information please contact our EU
representative  GPSR@taylorandfrancis.com
Taylor & Francis Verlag GmbH, Kaufingerstraße 24, 80331 München, Germany

9 780367 725983